THE COMPLETE IDIOT'S GUIDE® TO

Genealogy

Third Edition

by Christine Rose, CG, CGL, FASG, and Kay Germain Ingalls, CG

ALPHA

A member of Penguin Group (USA) Inc.

ALPHA BOOKS

Published by the Penguin Group

Penguin Group (USA) Inc., 375 Hudson Street, New York, New York 10014, USA • Penguin Group (Canada), 90 Eglinton Avenue East, Suite 700, Toronto, Ontario M4P 2Y3, Canada (a division of Pearson Penguin Canada Inc.) • Penguin Books Ltd., 80 Strand, London WC2R 0RL, England • Penguin Ireland, 25 St. Stephen's Green, Dublin 2, Ireland (a division of Penguin Books Ltd.) • Penguin Group (Australia), 250 Camberwell Road, Camberwell, Victoria 3124, Australia (a division of Pearson Australia Group Pty. Ltd.) • Penguin Books India Pvt. Ltd., 11 Community Centre, Panchsheel Park, New Delhi—110 017, India • Penguin Group (NZ), 67 Apollo Drive, Rosedale, North Shore, Auckland 1311, New Zealand (a division of Pearson New Zealand Ltd.) • Penguin Books (South Africa) (Pty.) Ltd., 24 Sturdee Avenue, Rosebank, Johannesburg 2196, South Africa • Penguin Books Ltd., Registered Offices: 80 Strand, London WC2R 0RL, England

Copyright © 2012 by Christine Rose and Kay Germain Ingalls

International Standard Book Number: 978-1-61564-156-7
Library of Congress Catalog Card Number: 2011936774

14   13   12   8   7   6   5   4   3   2   1

Interpretation of the printing code: The rightmost number of the first series of numbers is the year of the book's printing; the rightmost number of the second series of numbers is the number of the book's printing. For example, a printing code of 12-1 shows that the first printing occurred in 2012.

Printed in the United States of America

Most Alpha books are available at special quantity discounts for bulk purchases for sales promotions, premiums, fund-raising, or educational use. Special books, or book excerpts, can also be created to fit specific needs.

For details, write: Special Markets, Alpha Books, 375 Hudson Street, New York, NY 10014.

Publisher: *Marie Butler-Knight*

Associate Publisher: *Mike Sanders*

Executive Managing Editor: *Billy Fields*

Executive Acquisitions Editor: *Lori Cates Hand*

Development Editor: *Megan Douglass*

Senior Production Editor: *Janette Lynn*

Copy Editor: *Christine Hackerd*

Cover Designer: *Rebecca Batchelor*

Book Designers: *William Thomas, Rebecca Batchelor*

Indexer: *Brad Herriman*

Layout: *Brian Massey*

Proofreader: *John Etchison*

Contents

Appendixes

Foreword

You've been curious for quite a while. You've asked some questions of other members of your family, maybe poked around in the family papers, or even searched for others of your name on the Internet. But now you're serious about tracing your ancestors. However, just because you're serious doesn't mean you know how to tackle what seems like an unusual research project. That's why this book was written, to provide the background knowledge and skills necessary for successful genealogical digging.

Christine Rose and Kay Ingalls are both experienced genealogical researchers and teachers. They have done successfully for many years what they are teaching you to do—they interviewed relatives, pored through volume after volume in libraries, poked through old or abandoned cemeteries looking for elusive tombstones, wrote letter after letter, searched in courthouses and archives across the country. Throughout this book they identify pitfalls to avoid and tips to assist your research—the fruits of their accumulated research wisdom.

Genealogical research is challenging, but it is not necessarily difficult. Like any acquired skill, it takes learning and practice. Every genealogical researcher, no matter how skilled or experienced, started out like you; with a little knowledge about the family, a few records, and a consuming curiosity. Don't feel that you have to know everything at the beginning. Read through the book once to start, but come back to it again as you progress in your research. It will be there for you as you tackle new record sources or new steps in your search.

As your research progresses, you will likely discover what many genealogists eventually realize—the searching is almost as much fun as the finding! You will find yourself making time to "do genealogy." Supper can wait so you can have just another half hour in the library; you give up watching television to stare at computer screens; or you realize that since your family vacation plans take you so close anyway, maybe you can fit in a half day at a distant courthouse. And don't forget to share what you've found with your family. One find shared may stimulate further clues to investigate.

You will make discoveries along the way—and not just genealogical discoveries. You will learn more about yourself and the members of your immediate family. You will find ancestors who led quiet, ordinary lives. You may find scandals and secrets, or things our ancestors thought were scandalous but today hardly raise an eyebrow. You may find medical history details or hereditary factors that affect you and your family today. And you will find history much more interesting. Your view of the Battle of Gettysburg, the immigrant experience at Ellis Island or its predecessors, or the Salem witch trials is enhanced by the knowledge that your ancestors were participants.

Your ancestors are patient. However much time you put into searching for them, a couple of days vacation, one weekend a month, a few minutes whenever you can spare them, there will always be more you can do—another letter or e-mail to write, another microfilm to scan, another website to investigate, another book to check, another courthouse or archive to scour for clues. And when you do identify your fifth great-grandparents on your father's mother's side, each person in that distant generation had two parents waiting patiently for you to find them. You'll never run out.

—James L. Hansen, FASG

Since 1974, James L. Hansen, FASG, has been the reference librarian and genealogical specialist at the Library of the Wisconsin Historical Society, where he assists thousands of researchers a year. He is a nationally known speaker, having lectured at numerous conferences and seminars in the United States and Canada. Among his publications are articles on frontier families, early Wisconsin records, and Wisconsin newspapers. He was the 1994–1995 president of the Association of Professional Genealogists, and is a fellow of the American Society of Genealogists.

Introduction

We hope you'll be swept along with us in our fascination with family history. As a little girl, one of us spent many weekends at the home of grandparents. Creeping out of bed and perching at the top of the stairs, she spent hours listening to the elders recount the family stories. "But why didn't Dad, who had grown up in the same tiny town, ever discuss his own mother and her family?" she wondered. Her curiosity was aroused, and it simmered for years. During these same years, the other one of us, sparked by family memorabilia, was consumed with a desire to know the people who shared bits of their lives in their diaries. Who played with the marbles in the old-fashioned box, and who told stories of the bear whose tooth is still preserved? Who were these people? What were they really like?

Driven by our individual interests, we each picked up the same guidebook to genealogy and set upon a path that changed us forever. We didn't know each other then—in fact, we lived in different states. But we shared this passion for poking around courthouses, for solving the many puzzles we encountered, and for walking the ground our families walked before us.

During the years of research, we each made mistakes and learned from them. We read books, listened to countless lectures, and worked in the field. Then each of us, never losing the thrill of working with the dusty old records, obtained our certification from the Board for Certification of Genealogists and worked on the families of others, too. We still didn't know each other, but a move by one eventually brought us together in neighboring cities, and we began working as colleagues and enjoying a special friendship.

Several years ago we had a rare opportunity to introduce you to the adventure and challenge of genealogy, and to share with you some of the knowledge we acquired. Soon, we think, every minute you can spare will be spent delving through the treasure troves of records left by your family long before you were born. You'll enjoy the exhilaration of discovering that first document mentioning your ancestors, and then eagerly looking for the next, and then the next. You'll be hooked.

Though you can't do all your genealogical research on the Internet, it's one more valuable tool to use while on the trail of your ancestors. For this updated edition of our book, we devoted a chapter to the Internet, and we added tips and explanations throughout the entire book to jump-start your use of online resources. You'll find numerous website addresses within these chapters.

With the guidance on these pages, you'll learn to use the Internet to enhance your family history. At the same time, we show you how to go beyond the Internet to find and experience the thrill of holding a fragile and aged document—a document that not only tells something about your ancestors, but also may be the same piece of paper one of them actually held. Your family tree will take shape as you uncover stories that surprise, inspire, and enthrall you. History will come alive. Your ancestors will become real. Records that include height, weight, complexion, and eye and hair color, all enable you to visualize them. You'll learn if they were shoemakers or lawyers or farmers. You'll follow their moves as the frontier opened, and marvel at their initiative (and their anxiety) as they trekked west with their large families. If your heritage is African American, you'll find a whole new chapter devoted to the research of African American families. The pages in this book will give you all the basics to launch your own adventure, and include advanced ideas to apply as you develop expertise and a lifelong interest.

How This Book Is Organized

In this book, we take you step by step through the process of genealogical research. You'll master the techniques and learn about resources to track down the information on your family's unique history.

Part 1, Who Are You? gives you a glimpse of the excitement that awaits you in genealogy. Learn how to begin gathering the information you need to start your search and how to keep track of what you've found.

Your ancestors left a trail, and in **Part 2, Finding the Trail,** we teach you how to pick up that trail. You'll learn to track the correct surnames and to connect with distant relatives. You'll start looking for evidence of your ancestors' existence in their hometowns, in libraries, and—yes—on the Internet.

Part 3, Following the Trail, guides you in the use of one of the most basic tools for finding your ancestors: the census. You'll learn to research effectively from home via the Internet and by corresponding with individuals. You'll also get detailed instructions for using those records online or at repositories.

Part 4, In Your Ancestors' Footsteps, takes you on the road to the exact places where your ancestors made their own history. We practically tell you how to pack your suitcase! (If you can't travel now, don't skip this—the information here helps you even if you rely on searching from home.) We'll introduce you to the wonders waiting for you in courthouses and the enlightenment you'll experience while digging in cemeteries. Thought you knew how to read a newspaper? We'll show you new ways. Some surprises are in store: you'll learn how wartime service provides peacetime data.

The paper mountain that grows from your research will be tamed in **Part 5, Making Sense of It All.** This includes the sound practice of citing your sources and some techniques for writing your family history. You don't want just the dry dates of your ancestors' existence. You want to wrap them in the history of their times, so they come alive to you and others. We'll help you over, under, around, or through the brick walls that every researcher hits.

Part 6, Expanding Your Horizon, offers tips on getting the most for your money, and opens the door a crack to some advanced research opportunities, concluding with some advice to keep in mind as you follow the crooked paths to knowledge of the past. New to this edition is an expanded DNA chapter featuring the dynamic use of new tools for genealogists, DNA tests. We explain what it's all about and how you can use it effectively.

The appendixes include suggested resources and a glossary. Books and programs mentioned throughout the chapters are fully cited in the appendixes. There are worksheets to keep you on track, and census forms to assist with your census search.

Extras

Within each chapter are some special boxed notes to call attention to things that will help you:

DEFINITION

Here are the definitions of terms common to genealogy.

PEDIGREE PITFALLS

These sidebars offer cautions to help you avoid common mistakes.

TREE TIPS

These boxes contain genealogy gems to help your research.

LINEAGE LESSONS

Here you'll find notes covering extra information to enhance your study of genealogy.

Acknowledgments

Many thanks to the staff of Penguin (listed at the beginning of this book) who assisted us in so many ways—to Lori Cates Hand, executive acquisitions editor, who was responsible for our updating this book, to Megan Douglass, our development editor, and to Janette Lynn, the senior production editor. We couldn't have brought this edition to you without their assistance. To all the others, the illustrators and the production team, our thanks, too. Though we didn't work with them directly, each had a hand in producing this book. You would not be reading it without this dedicated staff.

Our thanks, too, to David Brown and Marcia Brown (no relationship to each other!), project administrators of one of the largest Y-DNA projects in the United States, whose time and knowledge were instrumental in the preparation of Chapter 24 on DNA. Others who assisted with input on DNA were Mary Fern Souder and Dr. Thomas Shawker. We sincerely thank them. To Michael Hait Jr., CG, our thanks for all his knowledge displayed in Chapter 12 on African American research. To Elissa Scalisse Powell, CG, CGL, our thanks for sharing her expertise in Chapter 5 on Internet research. To James L. Hansen, FASG, our thanks for writing the foreword. His vast knowledge in the field makes him sought after in many capacities. And to our colleagues who helped with and are named in the first edition, our continuing appreciation for their prior assistance. You all helped make this a valuable edition.

And we again thank our husbands, Seymour Rose and Don Ingalls, for their enthusiasm in embracing the first edition, then the second, and now this third.

Trademarks

All terms mentioned in this book that are known to be or are suspected of being trademarks or service marks have been appropriately capitalized. Alpha Books and Penguin Group (USA) Inc. cannot attest to the accuracy of this information. Use of a term in this book should not be regarded as affecting the validity of any trademark or service mark.

Who Are You?

What's the fascination? Curious as to why Grandpa never spoke about his family? Yearn to know your ethnic roots? Whatever it is, this part will get you going. Starting the search with your own family, you'll learn how to spot the significance of all the papers and memorabilia you're sure to find. If you've already started with the Internet, find out how to build on that.

You'll also get the basics of recording what you find and an introduction to some of the charts, forms, and logs that will help you keep on track. You are laying the groundwork for a wonderful adventure.

Why Genealogy?

In This Chapter

- Discovering your family's interesting history
- Reaping the benefits of exploring your genealogy
- Expecting the unexpected

Your friend has been so excited about his upcoming family reunion and the chart showing he's descended from Pocahontas, and it occurs to you that you don't know of any interesting historical figures in your family's past. While completing your child's baby book, you realize how little you know of your family. Perhaps a recent PBS series on ancestors made you wish you'd quizzed Aunt Mabel about your French-Canadian antecedents before she died. Maybe NBC's hit show *Who Do You Think You Are?*, which depicts the searches for ancestors, prompted you to consider your heritage. Or maybe a friend told you she found a website devoted to your surname. Whatever the reason for your newfound curiosity, you now long to know about the people whose bloodlines you share. Is it possible to find your roots?

What's the Fuss About?

Genealogy is said to be the third (some say the second) largest hobby in the country—just ask any librarian. Librarians will tell you their shelves are bursting with genealogy books. Library computers are continually used by those seeking clues to their lineages. The fascination with genealogy is hard to describe. Some enjoy putting the pieces of the puzzle together. Some love history. If your great-grandfather served at the Battle of Gettysburg, you connect instantly to that historical battle site. If Great-Aunt Peggy was the first white child born in Monroe, Indiana, then you establish a bond with that town. Some are curious about a family story. Was your grandfather really expelled from college because he and some others hoisted a cow to the top of the belfry?

DEFINITION

Webster's tells us **genealogy** is the account or history of a descent of a person, or a study of a person's family. The hobbyist will tell you genealogy is a madness—an addiction that will forever change how you spend your every spare moment.

Some embark upon this journey because they need to track their family's medical history. The diagnosis of a disease or congenital condition may induce a descendant to document the condition as it was passed down in the family. Another budding genealogist may be interested in a lineage society, such as the Society of the Colonial Dames or the Sons of the American Revolution. Whatever the reason, all who begin the journey of tracing their ancestry share a common opinion. It is addictive. All your extra time is spent writing relatives, searching through documents, and researching online. Happily, this addiction has many rewards: it allows you to connect with new-found relatives, links you with new friends in every part of the country, and reveals fascinating bits of history and folklore that enrich your life.

As you travel this new road you will discover resources you never knew existed. All states have at least one major genealogy *repository*, and some have several. Your state's repository may be located in the state library or the state archives, or there may even be a special state library specifically for genealogy.

DEFINITION

A **repository** is a physical location where items are placed for safekeeping. This may be a museum, library, archives, courthouse, or other similar place.

If you are a newcomer to genealogy, you will be amazed at the variety of information and the multitude of sources. Your first trip to a genealogy library or to a National Archives branch may be overwhelming. "I can't believe there is really an entire book written about my family," you will exclaim when you find a genealogy published in 1875. As you begin to realize all the information you can find online, you'll spend more time at the computer. You'll be excited when you find your grandmother and grandfather in 1930 census records, and you see your mother listed as a small child in their home. To your surprise, you also note an uncle you never heard about. What joy! You may even acquire an obituary about a relative or a biography that includes a photo of a family member, either in a published book or online.

LINEAGE LESSONS

There are 13 branches of the National Archives spread through several states, in addition to the main Archives I in downtown Washington, D.C., and Archives II in College Park, Maryland.

Why the Effort?

With all the entertainment available, why would you want to spend your time writing letters, sending emails, talking to relatives, and visiting libraries and cemeteries? A definitive answer defies all who attempt to explain. For many, tracking one's family history provides a sense of identity. Those who believe their family is too ordinary or too poor to be interesting are often amazed to realize the sense of value they place on their family and heritage after delving into the family background. Tales of hardship and endurance can instill within you an understanding of the conditions under which your ancestors lived. My husband's grandfather spoke proudly of how his family was too poor to buy fruit, and how his mother grew tomatoes and learned to make a wonderful sweet green tomato pie. The courage and resourcefulness of your ancestors will help you establish a bond with them, no matter what their stations in life. You will feel their pain when they lost little babies, and when the older children had to drop out of school because they had no shoes to wear.

These are among the many reasons for interest in genealogy:

- To determine ethnic origin
- To explain why your dad wouldn't talk about his family
- To find out if you are *really* descended from a significant historical figure
- To indulge a passion for history
- To note traits in the family, such as temperament and talent
- To satisfy the need for a sense of identity
- To track congenital health problems
- To join a hereditary society

- To track down a family tradition

- To identify migratory patterns

- To discover original owners of family artifacts

- To determine ancestors in a particular occupation

- To reclaim the family cemetery

Will It Really Grab Your Attention?

You have your own reasons for your interest in genealogy. However, though the spark is there, you still wonder whether it is going to be worth the effort. "Why bother?" you think. "My ancestors are all gone. What difference does it make?"

It is hard to recognize, this early in your quest, that your search will affect positively not only your immediate family, but others as well. But, in a short time, you'll experience the enjoyment it provides. When I developed my own interest in genealogy, my husband's grandmother was grateful that someone was interested enough not only to listen to all the stories, but to make sure the family mementos were not discarded: the locks of hair, the mother's pin worn during the war, and the old postcards that were exchanged, and very old letters written by family members. It brought her considerable peace to know that these items would be treasured and preserved by future generations. On another branch of our family tree, our interest generated a family reunion, and relatives who hadn't seen each other for years were once again able to connect. An interest in family history touches many lives.

TREE TIPS

Ask your family members if there is an existing genealogy book or chart within the family. Someone may have already worked on the family tree, and it will give you a wonderful start.

Starting Down the Road

My own genealogy adventure started around 1960. A small act was destined to shape the years that followed. My husband's maternal grandfather visited and brought with him some family papers. He thought we might like to look at them. We did, and looking at those papers changed the whole course for this family! I decided to write down the names of the grandparents and their parents, "just in case the children ever become interested." My curiosity was immediately piqued when I noted that some of the family members' spouses were unknown to us. "Who were *they?*" I wondered.

The Path to Addiction

I went to the local library and borrowed a little book titled *Searching for Your Ancestors*. I read it cover to cover. Our children were still young, so a hobby away from home was impractical. However, here was something I could do at home in my spare time with no timetable. I could do as little or as much as I pleased, and whenever I pleased. My work had no deadlines to meet—I just wrote letters when I could and went to the library when I wanted. It was ideal! Looking into our family's past also appeased my interest in history. While tracing our ancestors' migrations from the East to the pioneer West, their service in various wars, and their encounters with Indians and dry prairies, history came alive for me. I could (and did) spend hours going through family memorabilia and documents, becoming acquainted with each ancestor. The daily mail was the highlight of my days, as I eagerly sought answers from county clerks, National Archives, fellow researchers, and relatives. Computers were in the future when I started, but I used all the traditional methods to carry out my search. I wore out many typewriter ribbons writing to repositories, potential relatives, newspapers, and anyplace and anyone else I could think of.

LINEAGE LESSONS

A maternal ancestor is an ancestor from the mother's side of the family; a paternal ancestor is from the father's side of the family. Your father's mother is your paternal grandmother; your mother's mother is your maternal grandmother.

Who Is an Ancestor?

Perhaps by now you are wondering which of your many relatives are considered "ancestors." Whose bloodlines do you share? Your ancestors are those from whom you are descended directly. The term "ancestor" usually refers to someone who existed prior to your grandparent. Your aunts, uncles, and cousins are relatives, but they aren't your ancestors.

Your great-grandfather is an ancestor; his brother is related to you, but he isn't your ancestor. You have two parents, four grandparents, eight great-grandparents, sixteen great-great-grandparents, and so on. By the time you trace back 10 generations or approximately 300 to 350 years, you have 1,024 ancestors. This impressive figure is more than enough to keep any researcher busy for a lifetime.

The following ancestor diagram starts with "You" (left), and then moves backward for three generations. The father's line moves toward the top of the chart, with the mother's line moves toward the bottom. See Chapter 4 for another example of a pedigree chart.

What Family to Trace?

You will, as you progress, need to decide which lines are of the most interest to you. Otherwise, you'll become overwhelmed with information. Decide whether you'll be content to trace the lines that lead to an immigrant ancestor who arrived in America, or whether you'll want to pursue certain ancestral lines in their countries of origin.

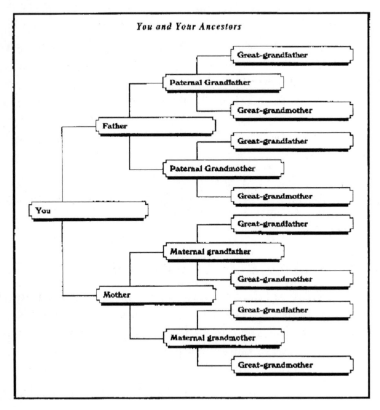

Ancestor diagram.

The generations closest to you are the family names that first appeal to you. The name you bore at birth (your father's line), your mother's family, and your paternal and maternal grandparents' families are familiar names that will influence your choice. As your search continues, other names will capture your attention. Perhaps the line that produced artists will fascinate you, especially if your child has some talent in that direction. The family that left their comfortable surroundings in Philadelphia and moved to the "wilderness" of Ohio bringing only meager possessions may stimulate your imagination.

What Will You Get from Genealogy?

Ask 10 people what they get from genealogy, and the answers will vary widely. A distraught mother is sure to respond with "A messy house! Files overflowing!" A young father may say "the poor house," as extra funds go for books, documents, and subscriptions to online databases. A wife will sigh as her husband dashes to the computer every evening.

But what are the rewards? Meeting relatives you never knew. Making friends from all over the country. A sense of completeness in putting names to your family members. Identifying with your ethnic background and someday visiting your country of origin. Satisfy a passion for history. The list is endless. The rewards come not only to you as the researcher, but also to all those you touch during your exploration. Relatives will be forever grateful for your efforts and will embrace you eagerly when you meet.

Your interest will spark contacts beyond your own. Your children will become acquainted with relatives they never knew existed and visit towns and areas they wouldn't otherwise have seen.

Embarking on the Adventure

When starting our own search, I found that my husband's paternal southern line was a real challenge. His great-grandfather had crossed the plains to California in 1856. My husband's grandfather was the youngest sibling, and he was only four when his father died in 1879 in the Sierra Mountains, still searching for gold. The tooth he saved from the bear he killed in the Sierra Mountains, shown in the following figure, serves as a reminder to his descendants of the frontier dangers he faced.

My husband's mother recounted a few stories, but she didn't know her husband's family, and his grandfather knew only that his father was from Alabama. The search seemed insurmountable. Then a family Bible turned up, and there was listed the county of his birth. I wrote some letters to other researchers and located that branch of the family. What excitement we felt to finally find these relatives!

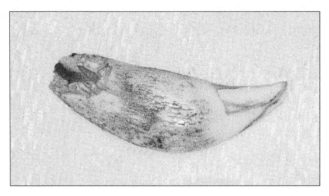

My husband's paternal great-grandfather saved this bear tooth from a bear he killed in the Sierra Mountains.

TREE TIPS

When meeting a large group of your family for the first time, ask each to sign a *Trip Memories* book for you. This book needs to include your family members' full names, addresses, and, if they have one, their email addresses. Ensuring that they provide this information will allow you to get in touch with them later. If you ask that they also add the names of their parents, it will also help you identify their branches as you become familiar with the family.

About four years later, we packed up our four small children in a tiny trailer and undertook a trip to their state with a great deal of trepidation. We didn't know what to expect. At that time, it had been 112 years since there had been any personal contact between the families. Happily, we arrived to a warm and loving reception. Amid a huge family reunion, we met hundreds of relatives. We will always remember the family members in their 80s (first cousins of my husband's grandfather) telling the stories they'd heard about the final goodbyes when their uncle left for the west, never to be seen again, and the tears in their eyes when they met my husband, their first link to our branch. Everyone was wonderful.

While we visited an old farmhouse with its big front porch, large yard, and cornfields across the road, our relatives told us the stories we yearned to hear. They took us to the cemeteries where members are buried, introduced us to other relatives, and treated us as their own. Some family members even had us in their homes. This initial road trip was only the first of many visits. While going back to Alabama, walking the grounds on which the ancestors lived, and visiting their final resting places, we felt like we were a part of their lives. Until then, we had been isolated from this branch of the family. Why genealogy? Because the joys of discovery are endless.

You Won't Find Perfect People

The results of your search may surprise you. I had been told many times that one of our relatives was a famous figure in the Revolutionary War. Indeed he was—on the other side! The supposed tie to a president may prove unfounded when you accumulate the facts. Grandpa's uncle perhaps never died young, as the family said, but instead was jailed and branded for stealing a horse, and the humiliated family never spoke of him again.

PEDIGREE PITFALLS

One common problem you will encounter in family stories is that many assumed incorrectly that they're related to others who possess the same surname. If the name was Adams, they assume a relationship to the two presidents with the same name, even if there's no evidence of a familial connection. Similarly, a family of the name Lincoln who claims a family link to President Lincoln may one day discover the purported link is false.

Be prepared. Know from the beginning that you may not find what you expect. Be charitable and nonjudgmental. Your ancestors made mistakes; they had weaknesses and strengths; they were human. You aren't embarking on the research path to find perfect people. You're seeking knowledge of their way of life and what made up the fabric of their character. The remarkable musical skill of your youngest child may be explained when you learn that three family members were fiddling in the Blue Ridge Mountains at the ages of four, five, and six. There will be disappointments. Your hopes of joining the Daughters of the American Revolution may disappear when you can't find any ancestors who participated in that war effort. Whatever disappointments the search may hold, they'll be replaced hundredfold by understanding of their lives, and enjoying the friendships you make while on your search.

LINEAGE LESSONS

Many middle and secondary schools now offer school genealogy projects that focus on tracing a physical trait (biology), improving letter-writing and email-writing skills (English), or enhancing other study courses. Even using a computer for research can be a skill honed by a genealogy project.

Genealogy with Success

There are no guarantees that you'll be able to trace your lines successfully. Certainly, out of the 1,024 direct ancestors in 10 generations, there are likely some lines you can trace back to their entry into North America. Be patient and follow all clues. Those who achieve the most success are those who constantly follow all leads: church records, obituaries, vital records, military files, and census data—use everything that could include records of your ancestors. Successful genealogists constantly restudy existing material as new data is found, to glean clues they may have missed previously. If you're systematic and use sound research techniques, you'll be rewarded. No one is pushing. If you need to take two years off from the search to complete your college education, no one is demanding that you write emails investigating your family's genealogy. When you're ready, genealogy is there for you to pursue.

No Two the Same

Remember that every life is different. Each life is important. Each life had its joys and its sorrows. The stories of courtship and of moves by teams of oxen and early rail will enthrall you. When I learned that one great-grandfather left New York as a small boy and sailed the Great Lakes to Wisconsin, it led to a study of the great sailing era, when hundreds of boats dotted the lakes. Finding that a great-great-grandfather was a "forty-niner" to California and had sailed around the Horn was almost as much fun to study as learning that a couple of years later his young daughter followed him from the East, by way of the treacherous Isthmus of Panama crossing.

TREE TIPS

Periodically reexamine the material you've accumulated. Today the name of "Elias Jenkins" in an old letter may mean nothing, but three months from now you may realize that had to be a son-in-law's name.

The Past Has a Personality

You will find opportunities at every step of the search to bring history alive for you and your family. The search isn't about collecting names; it's about identifying, with certainty, each of your ancestors and learning enough about their lives to forge a connection. When you read a will written in 1715 and realize what few possessions they had and how they parceled them out, you'll understand their lives of bare necessities. When you find the 1850 inventory of an estate that lists a shoemaker's tools, you'll learn your shoemaker grandfather was following in the family trade. A 1906 letter written from San Francisco will give you a jolt when you realize that was the year of the big earthquake and fire. Opportunities to know your ancestors are endless. Enjoy them at every step of the search.

The Least You Need to Know

- Genealogy makes no demands; you're free to invest as little or as much time as you want.
- There are numerous reasons to trace a family's genealogy.
- You can follow a family's trails in various ways.
- Your ancestors are those from whom you descend, the ones whose bloodlines you share.
- The only guarantee in genealogy is that you'll become totally absorbed in this fascinating pursuit.

Start with Mom and Dad

In This Chapter

- Getting started
- Interviewing relatives
- Finding clues among the family mementos

A notebook and pencil are all you need to get started. This is one of genealogy's major appeals—it requires only the basics, plus an inquiring mind and a sense of adventure. A tape recorder, though not essential, helps immeasurably to preserve the stories you hear. Someday, the family members you interview will be gone, and you'll treasure the sound of their voices telling tales of the past.

You may find it best to keep track of what you find in a word-processing program. If you wish, there are commercial programs available for purchase that allow you to enter your data as you find it.

You'll learn about genealogy charts in Chapter 4. It's best, however, to avoid being confined by charts. If you choose not to use a computer program to track your data, be sure your handwritten charts don't restrict your ability to add notes. If you use a genealogy software program, pick one that allows you to include ample information and provides the ability to add photographs and documents at a later time.

Getting It Down on Paper

When you take notes, date the first page, document the name of the person you're interviewing, include your name as the person conducting the interview, and where the interview was conducted. You may think that you'll never forget, but you're almost sure to after 10 such interviews. Years later, when your grandchildren find the notes, they'll want to know these details.

TREE TIPS

It is especially important when interviewing relatives to avoid being encumbered by a form. Do, however, take a list of questions, prepared in advance, with you to ask of your relatives.

Keep your interview notes saved in a labeled folder on your computer or in a loose-leaf notebook. You will refer back to this often. Organize the notes into subfolders or tabs by family so that you can instantly find what you need. If you're documenting the interview by hand and then inputting your notes into an electronic file, transfer the notes to your computer immediately after the interview and organize them by family name for easy retrieval or in the appropriate place in your genealogical software. If you use a recorder you will have the added enjoyment of listening many times to your relative's voices. Your descendants will treasure it, too.

While you're querying your relatives for dates and locations of family events, ask about the personalities and physical appearances of your ancestors. You want to know whether your ancestors were kindhearted, quiet or boisterous, or big-framed with a flowing beard—not just when and where they were born.

When talking about a woman, always ask about her maiden name. Uncle John told you his grandmother was Martha Jackson when she married his grandfather. Later, you find her obituary and puzzle for weeks over why it shows she was born Martha Smith. You finally visit Uncle John again and ask him about the possible discrepancy. "But of course," he says, "She was born a Smith but was married first to Joseph Jackson before she married my grandfather." Ask the right questions!

Questions trigger recollections of incidents and people that older relatives haven't thought about for years. Give them time for those memories to return. Don't rush through your questions; encourage them slowly. You won't get all your answers in one day. If your sense that the family member is confused about a memory, and is getting flustered while trying to remember, move on to another subject. Later, on another visit, you can come back to that topic and see if your family member is able to remember.

If you traveled a long distance to visit elderly relatives, you may need to get as much information as you can during that visit. Proceed gently. If they're confused, turn to other questions. You can always return to a subject later.

PEDIGREE PITFALLS

If Aunt Hattie says "Grandma was a Scott before her marriage …" ask her *which* Grandma. It's easy for people (especially for older people) to switch sides of the family as they relate the tales, without making that switch clear.

Don't Be Too Pushy

Watch for signs that Grandma is reluctant to give details. Don't press. If she says, "I don't want to talk about it," let it go for now. There may be a family secret, and pushing too hard may silence her for good. Be patient. The secret may be nothing more than her grandfather's divorce, which was a scandalous event indeed in 1850. Don't ridicule or discount a family member's reluctance to discuss a certain topic. Some family rifts have existed for years; feelings run deep.

Remember that times have changed. Behavior that's tolerated today was shameful in the past. There may have been some deeply hurt feelings when your great-grandmother married a man half her age only two months after she was widowed. Some family members may have felt shame because Great-Uncle Al enjoyed playing cards on the riverboat. And heaven forbid if there was a shooting or jail term!

Obstacles such as family secrets can slow a search, but getting the information is often crucial to continued success. If you don't know that there was a second marriage, you may not know under what name your grandmother was buried when you try to find her tombstone. Be patient. Ask the same questions of a number of relatives; you're likely to get the needed information in time, although you may never get the full details from the family.

If you suspect a notorious scandal, check the local newspaper instead of pressing the family too hard for details. There are many digitized images scattered among various libraries and online subscription services.

TREE TIPS

Be sensitive. Let your family know you aren't trying to be nosy. Assure them you just want to know about your family because it's interesting to know your roots.

Don't Believe Every Story You Hear

As a child, it was fun to play the gossip game. Two lines were formed, a story was whispered into the ear of the first person in line, and that story traveled via whispers from ear to ear to the end of the line. The end person in each line recounted what he or she had been told, and the one closest to the beginning story was declared the winner. How often was the final story even close to the original tale? Even with only five or six people in each line, the original story was often unrecognizable. How can we then expect a story, or *family traditions*, to remain accurate after five or six generations?

DEFINITION

Family traditions are stories handed down from generation to generation, usually by word of mouth.

Be cautious of the tradition that the male immigrant came disguised as a little girl to hide from kidnappers, as an extraordinary number of families have similar stories. Another common tradition is that three immigrants came: one went north, one went south, and one went west. All family traditions should be carefully noted, but find documentation confirming them before you, as the genealogist, accept them. Don't ridicule family traditions, even when you suspect that they're false. Respect the feelings of those who believe them. You can correct them tactfully when you accumulate irrefutable evidence that proves the traditions incorrect.

As you learn the techniques of tracing your family, you can utilize and examine various documents to assist in determining whether a family tradition is correct. You can confirm or discard the tradition that your family was related to President Adams or Jesse James after a step-by-step process to prove the family's lineage.

LINEAGE LESSONS

Don't be confrontational if you know the information being told is incorrect. Memories can be faulty and age may be a factor. Still, take steps to verify all you learn from family members.

Questions to Ask Your Family

The best way to start your research is by asking lots of questions of your own immediate family. Don't just ask Mom and Dad; ask everyone: aunts, uncles, cousins, and especially older relatives. Write down all their answers in full detail. These questions give you a beginning point for the search. You need to know the names of those who you're seeking and some possible dates so you'll know which records to search, and—most importantly—the various locations in which they lived. Make a list of questions. You may want to include the following questions:

- What was Grandpa's full name? Nickname?

- When and where was he born?

- What was his father's name?

- What was his mother's maiden name?

- Did Grandpa have brothers and sisters? What were their names? When were they born? Where were they born?

- Where did Grandpa's brothers and sisters live? Did they always live there? Where else might they have lived?

- What were the names of Grandpa's aunts and uncles?

- What did Grandpa look like?

- Does anyone have any photos of Grandpa?

- What did Grandpa do to earn a living?

- Where is Grandpa buried? Is there a tombstone?

- Did Grandpa ever serve in a war?

- What church did Grandpa attend?

- Did Grandpa have a trade? A hobby?

- Did Grandpa own land?

- What was Grandpa's nationality?

These basic questions will get you started. Ask the same questions about your grandmother. The answers will lead you to more questions and to the records your ancestors left. Until you know approximate dates and locations to point you in the right direction, there's no track to follow.

Searchin' the Attics and the Basements

Remember those old trunks and boxes you saw when playing around the attic and basement as a child? When you asked about them, your mom said they were just some family things. Ask your mom to go through the attic with you. Make sure you have plenty of time to explore, and don't hurry her; she hasn't seen these things for a long time. As you and your mom lovingly remove items from boxes, memories will come flooding back to her. Have a pencil and paper to make notes. Get the stories behind the items if you can.

Have a recorder handy as you go through the boxes and trunks. Record descriptions and ask questions regarding the objects. Be sure to announce on the tape the date, location of the interview, your name, and whom you are interviewing. Leave it running; you don't want to be distracted by turning the recorder on and off.

Who owned the little doll? It was wrapped with such care—it had to be very important to a little girl at one time. The little toy soldiers at the bottom of the trunk are sure to stir memories of the little boys who played with them endlessly. Who were they?

A little china doll like this one, packed with loving care, can stir up a lot of memories with your interview subject.

TREE TIPS

Many scraps of old paper provide clues of some kind. What appears unimportant to you today can later lead to the answer you seek.

The Objects with Tales to Tell

What about the objects that are going to help tell you about your grandpa and those before him? There's a batch of old letters with a ribbon carefully tied around them, and they're still in good condition after all this time. You open each letter, and read them in amazement. The first is a letter from your great-great-grandpa, written in 1889, when he went to Texas to look for a piece of land. He writes to the family that he's getting discouraged, that he misses them, and that he's about to come home. But then, in another letter dated three weeks later, he notes that he finally found what he wants. He describes the land and the little farmhouse on it. He promises to return home soon to bring the family to Texas. Your great-great-grandpa notes that Uncle George has found a place, too, and that his family will live nearby. Uncle George's wife Mary is already starting to fix up their home. Now you have some names and locations that will prove important to your search.

Family members, when queried, will often say, "But I don't have anything that would help." It isn't that they're reluctant to assist; they just don't realize the significance of what's stored in those old boxes in their attics.

Among the family mementos to look for are the following:

- Photographs
- Bibles
- Documents (such as deeds and wills)
- Letters
- Applications to lineage societies
- Scrapbooks and news clippings
- Funeral cards
- Account books
- Diaries and journals
- Baby books
- Christmas lists and address books
- Greeting cards
- Needlepoint samplers
- School report cards and diplomas

This list includes only a few of the *memorabilia* that may hold clues to your family's history. Perhaps there's a yearbook, a letter from an alumni association, or an invitation to a school reunion. Be on the lookout for anything that may give you an idea of where you might find further information.

DEFINITION

Within a family, **memorabilia** are those items that hold significance to the family. This may be a first baby shoe or a wedding announcement, or any other items that evoke memories of the family.

Picturing the Past

Likenesses of your ancestors are treasures. They may be faded *daguerreotypes* or hardly discernable *tintypes*. But hopefully someone has included a note as to whose likeness appears in the image. (A good photo shop can restore old images with amazing results, as can software programs such as Adobe's Photoshop.) Many of the old black-and-white photographs are preserved remarkably well, especially if they haven't been subjected to light. Examine each and note any names and dates. Note the city of the photography studio where the photographer took the photo; such information can provide a location for those family members. Free online photo editing services are available at Photoscape, Picasa, and Picnik. Check them out and see what works best for your needs.

DEFINITION

Daguerreotypes are an early photographic process that produces images on light-sensitive, silver-coated metallic plates. **Tintypes** are made on iron plates varnished with a thin sensitized film.

When you visit relatives, take relevant photographs with you—they may help bring back your relatives' memories. Use a scanner to make digital copies that you can share and archive. The family will be thrilled that you've preserved their memories, and they may be able to identify some of the people in the photographs.

Sometimes it's possible to connect two branches of your family using old photographs. Your Ohio branch and your Missouri branch may have lost touch 75 years ago. If both have the same photo of the original family home in Ohio, the photo in common can help assure you each branch is from the same family. The photos belonging to your relatives can also assist you if one branch has an unidentified copy of a photo and another branch has an identified copy. Your photograph of an unidentified Civil War soldier may be the same photo in another branch that has a name and date marked on the back side.

The Old Family Bible

Have you ever really looked at the old family Bible? Take a good look now. Surprised to find that there is a section devoted to family records? Recording the family line in a Bible was not only common, but it was often the only written record of the births and marriages in a family. Examine the Bible carefully. Some of the old-style script can be difficult to read; the flourishes render capital letters especially hard to decipher, and it can be challenging to identify numbers correctly. However, as you improve your *transcribing* skills, you will be able to read many styles of handwriting.

DEFINITION

Transcribing is to faithfully duplicate the exact wording, spelling, and punctuation of the original.

Some websites have very good examples of old writing styles. The grandmommy of all genealogical websites, Cyndi's List (cyndislist.com), has a special category devoted to the subject. On Cyndi's List's homepage, click on Categories and then select Handwriting & Script from the list that appears. You'll even find old-style abbreviations for names and locations listed with links to websites with examples.

Get a photocopy of the Bible's pages, or take photographs (preferably using a digital camera) if it is too fragile to withstand the copy machine or scanner. Be sure to include an image of the title page to show when and where the Bible was published. "Why would I want to know that?" you wonder. It's important to establish when the entries were written. If the Bible was printed in 1850, but the first entry is dated 1775, then you know the entries were either copied from an older record or are based on some other source. Copying from another source means the information is subject to errors that occurred while the person was copying or transcribing the older record.

Knowing the publication date may help you resolve discrepancies when comparing the Bible with other documents. If the Bible was printed in 1850 and the first entry is a marriage in 1852, followed by births of children in the order they were born, the entries were probably made at the time of the event and may be more apt to be correct. Pay particular attention to the handwriting. Were all the entries in the same hand? Were some of the dates added with a ballpoint pen in a later, more modern hand? These observations will help you to evaluate the accuracy. Try to find out not only the name of the present owner, but also all the previous owners of the Bible.

TREE TIPS

Cyndi's List (cyndislist.com) is an absolute "must" for anyone doing genealogical research via the Internet. This site has over 200,000 links to information that may aid you in your quest.

"This Deed Dated the ..."

Scattered among those family papers you may find old *deeds*, *mortgages*, and perhaps even Army discharges. The names and locations they mention can reveal many clues. Other valuable documents might include a will that was never discarded after a new will was made by an ancestor, or a life-insurance policy with documents that include family background. Take careful notes while sorting through these papers, and list the documents so that you can refer to them in the future.

DEFINITION

A **deed** is a legal document used to transfer title; a **mortgage** is a pledge to repay money borrowed.

Letters: Speaking from the Grave

Faded and hard to read, old letters can capture a bit of your family's life. A letter written to a sister in 1855 states, "My wife Mary died and I have no one to help with the little ones ... Can you come and help for a while? ..." Written from California in 1850, a letter states, "We just arrived at the mines in Placerville, where the people are fighting for a spot to camp ..." pointing you to specific events, locations, and individuals. It will be frustrating when a letter is written to "Dear Sister" or "Dear Son," with no further identification of the recipient. However, as the search progresses, the recipient's identity may emerge. The names that may appear within the letter then become valuable new leads.

Also take note of the envelope that accompanies the letters: document to whom it was addressed, the manner in which it was addressed, and any dates that appear.

PEDIGREE PITFALLS

Beware of salutations such as "Cousin" Joe and "Aunt" Hattie. Relationships were often stated loosely. "Cousin" could really refer to a second-cousin or another relationship; an "aunt" may be a great-aunt or a step-aunt. Rarely did anyone include the "great" when addressing a relative. Also, watch for "Sr." and "Jr."; these don't always refer to father and son. They were often used in letters and in legal documents to distinguish between two people with the same name living in the same town. The two parties may be related, as uncle and nephew or in some other manner, or they may not be related at all. If they are related, the Jr. often becomes Sr. when the preceding Sr. dies. If the new Sr. has a son by the same name, the son (previously known as III) then becomes Jr. Watch for this switch.

"Ancient and Honorable ...": Lineage Societies

Joining lineage societies was very popular in the first half of the twentieth century and remains so. The societies are based on descent from veterans of various wars, from pioneers, from specific trades (such as tavern keepers), and many more. Watch for these applications. Information that applicants provide about their ancestors can assist in a genealogy search. Though most lineage societies' documentation requirements were looser in earlier years than they are now, the application still can provide valuable clues. Note the names of the *sponsors*. They knew the applicant and might be leads to further records.

DEFINITION

Sponsors are those who vouch for the suitability of the applicant who is requesting admission into the society.

Membership in these organizations has grown and flourished; currently there are hundreds of such groups. Some have published their members' lineage records, and most have websites. Go to cyndislist.com, click on Categories, and scroll down to and select Societies & Groups. Or if you know the name of the society, simply enter that name into your web browser. If you suspect that someone in your family joined such a society, obtain the society's address and request that family member's application.

In order to gain as many clues as possible from lineage papers, be sure to ask the organization for a copy of its membership requirements or check their website. Doing so may help you to understand the records connected with the application. Daughters of the American Revolution, for example, will admit descendants not only for the Revolutionary service of an ancestor, but also for the ancestor who provided supplies for the war effort.

Account Books: Not a Penny More

Account books, kept by the father (or head) of the family to record money transactions and other miscellaneous notes, often include the cash advances made to the children to purchase a farm, buy equipment, purchase household items, or for any other reason. These advances were usually recorded faithfully, to be settled at the time of the father's death if they were still due. When he made his will, a father often meticulously listed the cash advances down to the penny. He made sure that those advances were accounted for against the child's portion of the estate when he died. The account books may contain various other transactions: money put out to interest, implements purchased, and perhaps even family births, deaths, and baptisms. Scrutinize them carefully for clues to the writer's occupation.

Dear Diary

Did a member of the family travel west, take a train trip through seven states, or travel back to "the old country"? The traveler may have left a journal. A careful reading might reward you with the names of relatives visited on the trip and perhaps some interesting sidelights.

It was especially popular to write diaries during the Gold Rush to California and during the journeys west to Oregon, Wyoming, Colorado, and other points. The Civil War also produced numerous journals, though many did not survive. Those who were unable to write during the war period often put their recollections on paper after their return home.

Diaries can tell you a lot about the writer and his or her daily life. From the prairie, you hear about the heat, the dust, and the deaths. You also hear about the births, the fun the little children had, and the fear of being unexpectedly visited by Indians. Your ancestors' writings during the war tell you of the writer's loneliness, the fear, and the pain of losing comrades. But they also tell you of their hope for the future and of the pride in serving for a cause in which they believed. Your ancestors will come alive to you as you read their penned words.

Popular in the nineteenth century, charming autograph books contained poems, short writings, and eulogies. The following poem was inscribed at the bottom of one such book: "Selected for Belina Adams by her Grand Father in the 77th year of his age A Webster Lebanon Aug 30th 1828." Besides this book's genealogical value, there's some historic interest, because her grandfather A. [Abram] Webster of Lebanon, New York, was a brother of Noah Webster, of dictionary fame. Look among your family's papers and you, too, are bound to find such treasures.

How vain are all things here below
How false and yet how fair
Each pleasure has its poison too
And every sweet a snare
The brightest things below the sky
Give but a flatt'ring light
We should suspect some danger nigh
Where we possess delight
Our dearest joys and nearest friends
The partners of our blood
How they divide our wavering minds
And leave but half for God
The fondness of a creatures love
How strong it strikes the sense
Thither the warm affections move
Nor can we call them thence
Dear Saviour let thy beauties be
My souls eternal food
And grace command my heart away
From all created good

Selected for Belina Adams
by her Grand Father in the 77th Year of his age
A Webster
Lebanon Aug 30th 1828

This appears in an old autograph book of Belina Adams, daughter of Isaac Ward and Eunice (Webster) Adams.

Baby Books: A Mom's and Pop's Joy

Baby books, so lovingly written, are wonderful to read. You can sense the mom's excitement when the baby's first tooth poked out; you can feel the dad's pride swell when that baby's first steps were taken. Besides the list of gifts, which may name relatives, there may be notations: "He has deep blue eyes like Grandpa Smith" or "Everyone says she looks just like Aunt Margaret." Now you know there was an Aunt Margaret! Watch, too, for baptismal dates, new addresses, and other listings that point to more records.

TREE TIPS

Look also for engraved silverware. The initials may give a clue to a husband's or wife's name.

Address Lists, Samplers, and Other Treasures

Look for old address books, Christmas lists with addresses, and invitations to events such as a fiftieth wedding anniversary. Old greeting cards also are helpful in providing names and addresses, and family news. You want to find anything that may provide a lead to a relative or a town in which a family member lived.

Don't overlook the cross-stitch sampler. A popular pastime was to create one with the names and birth dates of all the family members. Friendship quilts created by a bride's friends as a wedding present may feature embroidered names in each square.

No Longer Junk

Those boxes and trunks have now taken on new meaning for you. No longer "junk," they're the means by which you're going to get to know your family members and the lives they led. It takes weeks to adequately search the memorabilia in your family, scattered among the aunts and uncles and grandparents. Once you have, you're ready for the next step in the path to tracing your family's roots.

Or Did You Start with the Internet?

You may be inwardly protesting, "But I started with the Internet, not with the family." That's okay. Just set the Internet information you've collected aside for now and, using the ideas in this chapter, start collecting answers from your family. It won't take you long to realize that there are a whole lot of new clues now that you're tapping into the family's recollections and their precious memorabilia. Many new ideas will emerge, giving you a fresh start. You'll be excited by all the leads to explore further on the World Wide Web.

The Least You Need to Know

- Start your genealogical search with your own family members—not only your parents, but also your aunts and uncles.
- Be sensitive; don't press family members who are reluctant to talk about an episode in the past.
- Perform a thoroughly exhaustive search of your family's mementos. Most will help your search in some way.
- The old family Bible often is the only written record of a family.
- If you started your search on the Internet, go back and do what's suggested in this chapter to expand your knowledge of the family.

You're Hooked— Now What?

In This Chapter

- Getting the right information into your notes
- Creating logs to help keep track
- Using transcription and abstraction
- Learning the language of old documents

As you gather the oddly shaped pieces of your family's puzzle, your excitement mounts. Eager to find the missing pieces and see how they fit into the picture, you want to rush ahead. But before you plunge into the wonderful world of records and documents just waiting to be discovered, pause for a few minutes and learn how to get the information you need from your research. This chapter and Chapter 4 describe some tools and techniques to assist you.

What Are You Looking For?

What are you trying to learn as you go through the family papers, interview family members, and research in libraries, archives, and online? For each ancestor, you want to know the following:

- Full name (and nickname when applicable)
- Date and place of birth
- Date and place of marriage
- Date and place of death
- Name of parents

These dry facts don't give much insight into your ancestors' existence—no hints at their joys, hardships, and relationships. You need these basics, however, to know where to look for the records that will help you to visualize your ancestors as individuals.

You read in Chapter 1 that your ancestor is one from whom you are descended. Your ancestors' siblings are related to you, but they are not your ancestors. Your ancestors' cousins are related to you, but they are not your ancestors. Nonetheless, it's important to learn as much as you can about these *collateral relatives* because they may lead you to information on your ancestors.

DEFINITION

A **collateral relative** is someone with whom you share a common ancestor but who isn't in your direct line. Your mother's brother is a collateral relative. Your grandfather's uncle is a collateral, too, as are your cousins, because you and they share a common ancestor.

It's important to collect the same information for your collateral relatives as you collected for your ancestors. It's especially important when you hit a brick wall. For example, I thought my great-great-grandfather, George Marvin, was the son of Sylvanus Marvin, but I couldn't find proof. There is no mention of George in Sylvanus's will, though several daughters (including Harriet Bush) are named. I started looking for the records of these collateral relatives, and eventually found Harriet's will. In it, she names her brothers and sisters (including George), and wills to them her share of her father Sylvanus Marvin's estate. Because of the record she left, I was able to connect another generation in my pedigree.

What Should You Use to Take Your Notes?

When you become interested in genealogy, you quickly accumulate pieces of paper with family stories and details. Even if you use a computer to keep track, much of what you do involves note-taking with paper and pencil.

Avoid the temptation to take notes on any handy piece of paper; the backs of envelopes, credit-card receipts, and odd-size note pads will get lost. Take your notes on loose-leaf paper, or in tablets or spiral notebooks. Loose-leaf paper is particularly advantageous because the sheets don't have ragged edges, and you can file your notes neatly under a variety of topics.

Develop a system of note-taking that works for you. I like to put the following information in the top right corner of each sheet of paper:

- The surname of the family

- The location where I was when I took the notes (for instance, someone's home, a specific library, a county courthouse, or online)

- The date I took the notes

TREE TIPS

Use only one side of each sheet of paper for note-taking. If you want to refer back to something included in a group of clipped-together sheets of paper, it's much easier to shuffle through them to find the required information if you don't need to flip each sheet over to look at a second side.

With these headings, I can tell at a glance what family I worked on, and when and where.

Use separate sheets of paper for different surnames, even if the information is from the same person. If your mother recites family stories about her mother and father, put the information on two sheets of paper—one with the father's family name and one with the mother's family name. It's easy to get so involved in note-taking that information pertaining to several surnames ends up on the same sheet. This can create a confusing situation when later you attempt to study each individual surname.

Whatever system you devise for your research, keep it simple and be consistent. However, be flexible enough to change your system if you read about or observe another approach you think will work better for you.

Always try to take notes in a manner that minimizes recopying. Each time you copy notes, the chance for error multiplies.

That Important Citation

After you label your notes, write a full *citation* for the source of the information you're going to be gathering. (For more on the correct format for citing sources, see Chapter 20.) You may get so excited about a great find that you forget to write down where you found it. You can't use undocumented information—that is, information with no source citation or an incomplete source citation—to prove your line.

You need complete source references for other reasons, too. In cases of conflicting information, you need to know the sources to properly evaluate which is most likely to be correct. You may later find that you neglected to get all the information the source offered and need to locate it again.

DEFINITION

A **citation** is the authority or source from which the information was taken, included to support facts. In genealogy, every fact needs at least one citation.

Writing It Right

After you have labeled your paper and have documented the source citation, you're ready to take notes. Although your notes should be complete, you can use standard and recognizable abbreviations, arrows, and symbols. They should be clear to you or to anyone reading your notes. Don't be too brief. It's better to have too much information than not enough. Something that seems inconsequential now may become crucial as your search progresses.

Write out the full names of individuals. If you know the middle name, include it. If a person has a nickname, indicate that by putting the nickname in quotes after his or her *given name*. Do not use parentheses to show the nickname. Parentheses between a given name and surname are used to enclose the maiden name, so it would be confusing to use it for nicknames, too.

DEFINITION

A **given name** is a person's first name, and the name given to a child at birth. It's sometimes referred to as one's Christian name.

What's in a Name?

If someone is known by more than one name, put the alternate name or names in parentheses after the surname, preceded by "a.k.a." (which stands for "also known as"). Here is an example: John Smith (a.k.a. John Taylor). The name change may have occurred, for instance, when John Smith was adopted by a Taylor and was known by both names.

Write down all of a person's known names. If he went by his middle name or by his initials only, note that information: Laurence William Holmes was known as Bill, Will, and Willie. It will be important one day for his descendants to know his multiple names because he may be listed under any of those names.

TREE TIPS

Here's a good rule to adopt: if you find anything in the record that seems amiss or unusual, note it. It may be the evidence that proves or disproves a link you're trying to establish.

Always note the spelling variations you find. They may be insignificant, a reflection of times when names were spelled phonetically, or they may be important, suggesting that you have information on two different individuals rather than one. (See Chapter 7 for more on spelling variations.)

For names that apply to both males and females (such as Gale–Gail, Gene–Jean, Marion–Marian, Frances–Francis, Leslie–Lesley), indicate whether the individual was a man or woman if you can determine that from the document. Doing so eliminates any potential confusion.

If you find an individual with a name usually given to someone of the opposite sex (remember the Johnny Cash song, "A Boy Named Sue"?), be sure to indicate that in your notes. A number of names once used for either sex have fallen into disuse for males: Eleanor, Mildred, Beverly, and Valentine were all used for males at some points in the past. The Social Security Administration put together interesting databases of popular baby names by sex from 1880 to the present: www.ssa.gov/OACT/babynames. Look there to check the popularity of the names you're working on—or the popularity of your own name in the decade of your birth.

Women and Their Changing Surnames

Women's names present a special problem. You may find women under their maiden (birth) or married name, or even under the name of a prior husband. Try to establish a woman's birth name in order to identify her parents because they're your ancestors, too. List her by her maiden name, and indicate the names of her husbands. In your notes, list Mary Jordan (her maiden name) and show that she was married first to John Jackson and then to Frank Swift. Her full name would be shown properly as Mary (Jordan) Jackson Swift, listing first her given name, followed by her maiden name in parenthesis, followed by the surnames of her subsequent husbands with the latest listed last.

LINEAGE LESSONS

Though the proper full name of a woman includes her maiden name and all married names, use the name she was known by at the time of the event when you refer to her in your narrative. If she was already married to Frank Swift when she and her husband moved to Indiana, you would say "Mary Swift and her husband Frank moved to Indiana."

When you're recording a female on charts, but don't have her maiden name, insert only her first (or given) name. In the preceding example, if you didn't know that Jordan was Mary's maiden name, then you would show her as Mary ( ). If you need to refer to her in your notes, show her as Mary ( ) Jackson Swift, using the blank parentheses to indicate that her maiden name is unknown. When you establish her maiden name, you can fill in the blank.

Be careful with women's surnames. The name you find in documents may be a maiden name or a married name. If a woman is widowed (or divorced) and remarries, the surname in the marriage record may be that of the previous husband. Sometimes this is distinguished by the record: "Mrs. Margaret Smith married Richard Carter" indicates that she was married previously to a Smith.

Place Names Can Be Tricky, Too

Place names should be fully identified by writing down the town, county, and state (or the equivalent divisions for foreign countries). These geographic divisions are important in genealogy because many of the records you need are in the towns and counties where your ancestors lived. Because many states have towns and counties of the same name, be sure your notes always indicate the state, too. To help identify counties when you have only the name of the town or city, try websites known as geographic name servers. The one located at http://geonames.usgs.gov will find and list all features with the name. For example, I chose a "domestic" (United States) search and entered "Los Gatos," the town where I was born. Twenty-nine "features" appeared in the search results, including cemeteries, the county in which Los Gatos is located, creeks within Los Gatos, and so on. When the town or city is known, you can locate the name of the county at http://resources.rootsweb.ancestry.com/cgi-bin/townco.cgi.

LINEAGE LESSONS

As your ancestors moved west, they often named the new area after their old home area. If you don't know their prior residence and they were pioneers to that area, the name of the new town might provide a clue. Among the original settlers of Granville, Licking County, Ohio, were people from Granville, Massachusetts. If your ancestor was one of the first settlers, the name of the new town, named in honor of the old, would be an important clue to a possible prior residence.

If the records you find mention landmarks or geographic features such as creeks or hills or roads, include them in your notes. They may help to distinguish between two different families in the area with the same surname. Abbreviations of place names (except for states) may cause confusion later, so avoid using them.

Dating Problems

When you insert dates in your notes, use the format of day, month (spelled out), and four-digit year: 10 January 1988. If you write the date 10/1/88 or 1/10/88, later you or others won't be sure if the date was January tenth or October first, and whether the year was 1888 or 1988.

Sometimes a record is unclear as to the date, or the date is included in two places in the document and there's a discrepancy. Be sure to include these discrepancies in your notes; they may be important later in your research.

Did You Miss Something?

A final word about your notes. Review them at the end of any research session or interview. Check to see if there's anything in your notes that isn't clear. You may not have access to that source again; be sure you have it right.

Beyond Notes: The Other Papers You Need

You will want to keep more than just your interview and research notes on paper. You'll also want to create some lists to keep track of the sources you check and the information you find when you do your research. You can create your own, use commercially printed forms, or download online forms from sites such as www. ancestry.com/trees/charts/researchcal.aspx (a research calendar) or www.cs.williams. edu/~bailey/genealogy (which includes a family record sheet, a pedigree chart, a pedigree fan chart, and several others). You can download these forms and use them to record family searches.

These lists, or logs, can help you organize your finds, decide on your next step, and eliminate duplication. Two kinds of logs are especially useful: the research calendar and the correspondence log. Another easy way to keep track of what you've found is to set up an Excel spreadsheet. After you've logged your findings, you can sort the columns by surname, or a number of other ways you might find useful.

The Research Calendar

The research calendar is labeled with the surname and problem you are researching and has at least four columns: the date you conducted the research, the repository where you conducted the research, a description of the records you searched, and a brief summary of your findings. You can add other columns (time period, library call number, and so on). The research calendar shows you at a glance what records (documents, films, or books) you've used. It isn't, however, the place to record your more expansive notes taken during your research.

You'll need a variation of the research calendar to keep track of your Internet research. Instead of noting documents, films, or books, list websites you've visited and their *URLs*. If you follow links from the initial website, include them in your list.

DEFINITION

URL is the acronym for uniform resource locator, the address of a page you view online. Without that address, you'll have difficulty returning to that page later. Because URLs often change, it's important to include the date you viewed the page in your research calendar.

Surname

RESEARCH CALENDAR

Search focus (brief statement of research problem)

Date	Repository/call no.	Source	Findings

Here's an example of a research calendar.

Correspondence Log

A correspondence log is a record of the letters you've written and the replies you've received. The log should have a designated space for the surname at the top and five columns with space for the date you sent the letter, the name of the person to whom you addressed the letter, a brief statement of the information you desire, the date you received the reply, and a brief note of the results. Some researchers also include a column to list any fees paid to get the information.

CORRESPONDENCE LOG				
Date Sent	**To Whom**	**Request**	**Reply Date**	**Results (Positive, Negative, Burned)**

Here's an example of a correspondence log.

Word by Word: Transcribing and Summarizing Documents

The research calendar and correspondence log summarize your research activity. In addition, you will have your notes, analyses, and photocopies of what you found. Your notes contain all the details that don't appear on the logs. These notes include *transcripts* and *abstracts* of documents. Transcribing and abstracting are particular kinds of note-taking.

DEFINITION

A **transcript** is a word-for-word copy of the text in a document. The transcriber changes nothing; everything is transcribed just as it appears in the original document—errors, punctuation, misspellings, and all.

An **abstract** is a summary of the text in a document, retaining all of the document's essential details.

Learning by Transcribing

Transcriptions are most useful when a document is very difficult to read (due to content or condition). It's also helpful to transcribe a document when you're unfamiliar with the type of document, or when you have a particularly complicated or unusual document. You may be forced to transcribe if the fragile condition of a document restricts photocopying.

When you begin your research, it's helpful to transcribe all the documents you find relating to your family. The practice will allow you to become familiar with old-style handwriting and will help you learn to recognize common phrases in similar documents. As your familiarity increases, you may decide to switch to abstracting the documents except in select cases.

You can hone your transcribing skills by visiting some websites. Although the Board for Certification of Genealogists is aimed at individuals who want to become certified or for consumers wanting to hire a genealogist, the site's "Test Your Skills" section is useful for all levels of researchers. Go to: www.bcgcertification.org and click on Test Your Skills. Print out two documents to transcribe and abstract. When you're finished, compare your results to the correct ones the site provides. Click on Skillbuilding: Your Learning Center to study a variety of lessons. A particularly interesting website for learning to transcribe is www.dohistory.org/diary/index.html. Here you can transcribe right on your screen and compare your transcription with theirs. An interesting feature on this site is the Magic Lens, which magnifies portions of the document as your mouse hovers over words. Practice reading the handwriting without the aid of the magnifier and then get instant feedback by checking your work with the magnifier.

TREE TIPS

Read a document several times before you begin to transcribe or abstract. Difficult words or handwriting may become clear after several readings, making it easier to take notes.

Abstracts: Summarizing the Document

For an abstract, you extract every detail that may shed light on your research. You're looking for *all* names, dates, places, and events. Examine documents carefully for unexpected information. A document may mention a person whose name doesn't mean anything now, but he or she may turn out to be a relative. Note any mention of a location; it may lead to other records.

Although an abstract is a summary that doesn't include every word and punctuation mark in the original document, it does include names and places just as they appear. If you think there's an error in a name or place (or anything else significant), you can include your correction or explanatory remarks in square brackets after the word. The brackets enable others to easily distinguish what was actually in the abstract, and what you added. It's permissible to correct simple words in the abstract; you may correct "funrale" to say "funeral." It's important, however, to emphasize the point that while you can correct some simple words and their spellings in an abstract, in a transcription you must retain all words and their spellings as they appear in the original document.

PEDIGREE PITFALLS

If the name appears as Jas., don't convert it to James. An abstract isn't the place for the interpretation. After you've examined all the records, you can then determine whether the person's name is actually James. What looked like an "a" in one document might turn out to be an "o" in multiple other documents, and Jos. is the abbreviation for Joseph.

The following figures show a will from an actual probate file, followed by the transcript and abstract of the document. Compare the transcript and the abstract to the original document and then to each other. The transcript includes every word from the original; the abstract is a summary of the important information.

This is the will of William A. Glass. Note the blotch and how it is treated in the transcription in the next figure.

La[st] Will & Testament of William A. Glass

Adams Ind. June 20th 1881.

I William A. Glass being of sound mind and memory do declare this to be my last Will

First: I Desire a desent Burial

Secondly: All of my just debts paid

Third All of my property both Real Estate and Personal property that I may own at the time of my death I will and Bequeath to my wife Mary Glass for ever for her to have the same in fee simple to have t[he?] wright to sell and convey the same in a[ny?] way that she may see proper

I also appoint my wife my Executor of my will and for her to settle the same in any way that she may see proper to save as much expense as possible

William A. Glass

Ezra L. Guthrie }

Joseph Dineger } Witnesseth

This is the brief will of William A. Glass, filed in Decatur County, Indiana.

Did you think the name in the will was Glap? The last two letters are the "tailed s." One could easily misread documents if the transcriber is unfamiliar with old handwriting, as with the tailed s. All spelling and punctuation has been retained in the example. When a word isn't entirely legible or is partly gone (as are two words in this document), the transcriber has noted so with brackets.

[Abstract of will of William A. Glass, Will Bk 5 p. 182, Decatur County, Indiana.]

Adams Co., Ind. June 20, 1881, William A. Glass, all property real and person to wife Mary Glass in fee simple. Appoints wife Executor. Signed: William A. Glass. Witnesses: Ezra L. Guthrie and Joseph Dineger. [If the copy shows the date it is proved in court by the witnesses, that would also be included in the abstract.]

Here's an abstract of the brief will of William A. Glass.

Although there are fill-in forms available for use in abstracting, they are constraining and I don't recommend them. Trying to conform to the order on the form means rearranging the information from the document. In attempting to adapt the information to fit the form, omissions can occur or clues in the wording and order might be lost. I prefer to shorten the original while retaining the order of information in the document as closely as possible.

Learning the techniques for transcribing and abstracting is essential; both are integral parts of genealogy research. Practice constantly, and examine various books of abstracts at the library. (See Chapter 15 for further discussion of abstracts.)

TREE TIPS

If you're ever in doubt about whether something belongs in your abstract, include it. If it turns out to be extraneous information, it doesn't matter. If it's something you later need and don't have, it matters a great deal.

It Says *What*?

Because you start your research with yourself and work backward, the first documents you're likely to research are more recent ones. These aren't usually too difficult to read because they're in a more modern hand or they're typed or electronic. However, as you move back in time in your research, you'll find many documents that are difficult to read because of terminology and handwriting style.

TREE TIPS

It may be helpful to enlarge the writing in a document using a photocopier, if possible, especially if the original document is in poor condition. Be sure the repository holding the documents allows photocopying before doing so.

You Can Read It, But What Does It Mean?

The terminology that baffles you may be complex legal phrases or obsolete clauses. It could just be a peculiarity in the writing of the individual who created the record. Working with the official documents of a specific place and time, you'll learn the terminology and recognize standardized words or sentences that aren't important to the interpretation of the document. But when you begin, if you transcribe the complete document as suggested (including all words and punctuation), it won't take you long to recognize the common statements that you can ignore. Until you become familiar with standard terminology or legal phrases in documents, read and copy everything in the document.

"Strange" Old Words

As you do your genealogy research, you'll likely encounter many unfamiliar words. Customs change, laws change, and word meanings change. Occupations become obsolete, ethnic groups assimilate, and new inventions require new words. Language is constantly changing through usage.

Sometimes you can determine the meaning of a word using the context of the words around it or through the interpretation of other documents. Sometimes the words will be defined in your regular dictionary. *The Oxford English Dictionary* is particularly useful because it includes many archaic terms. You can also consult a genealogical dictionary. An excellent resource is *What Did They Mean by That? A Dictionary of Historical Terms for Genealogists,* by Paul Drake, J.D. In addition to definitions, this book includes documents that may help you become familiar with the kinds of things you will encounter in your research. It reproduces two ledger sheets from the mid-1700s. You can practice your skills by reading the items in these documents and then looking at the compiler's lists to see if you interpreted them correctly.

Here are some useful websites for researching unfamiliar terminology:

- Archaic medical terms: www.genproxy.co.uk/old_medical_terms.htm.

- Old genealogy terms: http://genealogy.about.com/od/glossaries/Specialized_Dictionaries_for_Genealogists.htm

- Military terms: militaryterms.net.

- Old occupation terms: http://usgenwenb.org/research/occupatons.shtml

Find other useful sources for solving terminology mysteries using your web browser. In the search box, type the kind of terms you're seeking followed by the word "terms"; for example, "medical terms."

TREE TIPS

A common phrase may have variant meanings. "Meeting house" may refer to a New England town hall, or it may indicate a Quaker place of worship. The difference could affect your research drastically.

Latin and the Law for Genealogists

Legal documents and court records are full of Latin terms that you may need to define. It's important for you to know that when an index lists a name followed by *et al.*, this indicates there are other names connected with this document in addition to the indexed name; *et uxor* or *et ux.* means "and wife." The best source for legal terms is a law dictionary. Consult *Black's Law Dictionary* whenever you encounter legal terms you don't understand. In particular, consult the fourth edition of Black's for older terms. You may also consider consulting Bouvier's 1856 *Law Dictionary* at www.constitution.org/bouv/bouvier.htm.

LINEAGE LESSONS

The term *et al.* is the shortened form of *et alii*, a frequently used legal term meaning "and others." When you encounter it in an index, pursue the full record to determine the names of the additional individuals; often they are family members. Also seen frequently in an index is *et ux.*, shortened from *et uxor*, meaning "and wife"; *et vir* means "and husband." Examining the document closely will likely reveal her or his name.

Abbreviations

A number of abbreviations common in early documents are rare today. You will often see "inst" as an abbreviation for "instant." Knowing this will help you to establish a correct date if you know that "instant" indicates that the date referred to was in the same month as an aforementioned date. Thus, in a response on December 28 to a letter dated December 4, the letter writer may say, "In responding to your letter of the 4th instant," which is to say, "I'm replying to your letter dated December 4."

"Ultimate," in contrast to "instant," refers to the previous month. An obituary that appears in a newspaper on July 25 stating that the deceased died the 20 ult. indicates he or she died June 20.

For help in deciphering many of the abbreviations you'll encounter, see Kip Sperry's compilation *Abbreviations and Acronyms*.

Making Sense of Chicken Scratches

The handwriting in some documents may be nearly impossible to decipher at first glance. With practice and careful examination, however, you'll learn to read many styles. Even neat, clear handwriting will reflect the contemporary style or usage and can leave you wondering about some words. It helps to seek references with illustrations of letter combinations and writing styles. Consult *The Handwriting of American Records for a Period of 300 Years* by E. Kay Kirkham. In it are large illustrations of the manner in which letters of the alphabet have been written through the years, as well as the abbreviations for selected names. The book also has some transcriptions. You can attempt to transcribe the same documents and then check your work against the published transcriptions.

The chapter "Reading Handwritten Records" by Raymond A. Winslow, Jr., pp. 97-105, in *North Carolina Research 2nd edition* has a good illustrated discussion of peculiar letter forms, symbols, and abbreviations that you may see in old documents. A helpful website giving examples is http://amberskyline.com/treasuremaps/oldhand.html.

Now that you've learned a little about how to extract information from your research, go on to the next chapter to learn how to manage that information.

The Least You Need to Know

- You need to establish the name, and the approximate dates and places of births, marriages, and deaths, to start your search.
- When making notes, include complete details of where you got the information.
- Carefully note full names, nicknames, and variations.
- Transcribing documents will help you learn to read old-style handwriting and to understand the terminology.

Keeping Track of What You Find

In This Chapter

- Using charts to keep track
- Record keeping with computer genealogy programs
- Generating reports using your computer

Genealogy research is a little like the connect-the-dots pictures you did as a child; if you did not connect the dots in the right order, the emerging picture wasn't quite right.

To help connect your genealogy dots in the right order, you will use genealogy charts and family group sheets to record the research data. You can use these basic structures with pencil and paper or with a computer.

There are two general types of genealogy charts: ascendant and descendant charts. An ascendant chart starts with you; then, you fill in the blanks by moving back through the generations of all your ancestors. A descendant chart starts with an individual, and then you build the chart by moving through the generations of that individual's descendants.

These charts aren't considered a finished genealogy. They are research tools you can use to remind you of where you are in your genealogy research.

In addition to genealogy charts, you can use a family group sheet to keep track of individual families.

Pedigree or Family Tree Charts

The best chart to begin with is a *pedigree chart*, an ascendant chart. On this chart, you start with yourself and work back in time, generation by generation, filling in your parents, then your grandparents, then your great-grandparents, and so on, as far back as you can.

Think of the pedigree chart as a shorthand master outline of your bloodline. A quick glance at it alerts you to the blank spots in the information you've gathered, which helps you develop your research plan. When you notice that Great-Grandma Diana's maiden name is blank or that there's no marriage date or place for Grandpa Guy, you have a clear picture of the information you still need.

DEFINITION

A **pedigree chart** starts with you and shows the line of your direct ancestors. It is sometimes called a family tree, lineage, or ancestry chart.

The format of a pedigree chart is always the same: your name (or the individual whose ancestry you're tracing) is on the first line, your father's name (or the subject's father's name) is on the upper line, your mother's name is on the lower line. The upper track in a pedigree chart is the father's (paternal) line, and the lower track is the ancestral line for the mother's (maternal) line. You are number 1 on this chart. Your father is number 2 and your mother number 3. On a pedigree chart, the numbers for men are always even numbers, and the numbers for women are odd.

PEDIGREE PITFALLS

Pedigree charts don't allow space for source citation. Therefore, you shouldn't disseminate them unless you attach an accompanying sheet with full citations keyed to the information on the chart, or a family group sheet, which does have citations.

As you can see, you quickly run out of space for all your ancestors on a four-generation chart. To list the additional generations, you must create additional charts.

In numbering the pedigree chart, you are number 1 on chart 1. One of your great-grandfathers is number 8 on chart 1. Make a new chart to continue with your great-grandfather's ancestors. Each great-grandfather (who's number 8 on chart 1) is number 1 on chart 2. On chart 2, be sure to make a note to indicate he's also number 8 on chart 1, so you can easily follow the line in connected charts.

PEDIGREE CHART

CHART NO. _7_

Prepared by _____
Address _____
City and State _____ Zip _____
Date Prepared _____

No. 1 on this Chart is # _7_ on Chart # _1_

8. *Sylvanus Marvin* Con't on Chart #___
Born *1 Nov 1800* [Father of No. 4]
Where *Camillus, Onondaga Co., N.Y.*
Married When *ca 1817*
Died *14 Jun 1891*
Where *Onondaga Co., N.Y.*

4. *George Nelson Marvin* [Father of No.2]
Born *29 Jul 1818*
Where *Onondaga Co., N.Y.*
Married When *8 Apr 1839*
Died *29 Mar 1904*
Where *Grundy Co., Ill.*

9. *Ruth Rouse* Con't on Chart #___
Born *1791* [Mother of No. 4]
Where *Washington Co., N.Y.*
Died *before 1865*
Where *N.Y. ?*

2. *William Nelson Marvin* [Father of No. 1]
Born *2 Jan 1844*
Where *Onondaga Co., N.Y.*
Married When *1 Oct 1875*
Died *27 Jun 1904*
Where *Grundy Co., Ill.*

10. Con't on Chart #___
Born [Father of No. 5]
Where
Married When
Died
Where

5. *Marietta Hammond* [Mother of No. 2]
Born *4 Jun 1817*
Where *N.Y. ?*
Died *21 Oct 1889*
Where *Grundy Co., Ill.*

11. *Rebecca ( )* Con't on Chart #___
Born *19 Feb 1793* [Mother of No. 5]
Where *N.Y. ?*
Died *30 Nov 1862*
Where *Woodford Co., Ill.*

1. *Ida May Marvin*
Born *31 Jul 1888*
Where *Grundy Co., Ill.*
Married When *24 Mar 1908*
Died *15 Mar 1963*
Where *Bloomington, McLean Co., Ill.*

Clyde Edgar Dunn
Name of Husband or Wife

12. *Elias Drollinger* Con't on Chart #___
Born *15 Jul 1802* [Father of No. 6]
Where *Guilford, N.C.*
Married When *14 Sept 1823*
Died *2 Jan 1871*
Where *Wabash Co., Ind.*

6. *Mathias Drollinger* [Father of No. 3]
Born *30 Aug 1824*
Where *Preble Co. ?, Ohio*
Married When *29 Aug 1850*
Died *26 Dec 1895*
Where *Ind. ?*

13. *Annie S. Igo* Con't on Chart #___
Born *16 Dec 1800* [Mother of No. 6]
Where *Pa.*
Died *2 Aug 1874*
Where *Wabash Co., Ind.*

3. *Diana Cowell Drollinger* [Mother of No. 1]
Born *22 Feb 1857*
Where *Lagro, Wabash Co., Ind.*
Died *11 Jun 1946*
Where *Joliet, Will Co., Ill.*

14. *Simon B. Lloyd* Con't on Chart #___
Born *1797* [Father of No. 7]
Where *Va.*
Married When
Died *29 Aug 1855*
Where *Wabash Co., Ind.*

7. *Amelia Jane Lloyd* [Mother of No. 3]
Born *9 May 1835*
Where *Ind.*
Died *2 May 1886*
Where *Ill. ?*

15. *Mary ( )* Con't on Chart #___
Born [Mother of No. 7]
Where
Died *Bef 1842*
Where

A pedigree chart begins with a subject and works back through the generations.

Large or Small, They Chart Your Family

You can purchase blank pedigree charts accommodating anywhere from 4 to 15 generations. Four-generation pedigree charts are the most convenient to work with in your day-to-day research. The more generations on the chart, the less room there is for data about individuals.

Some decorative pedigree charts are suitable for framing. They can be fan shaped with the lines radiating out from you at the center. Some are in the form of a tree with limbs and branches representing family lines. Still others have spaces for photographs. Although attractive as wall art, these charts are too large to be useful as research aids.

Some charts are available free for personal use at various websites. One such site is http://genealogy.about.com/od/free_charts/ig/genealogy_charts.

Filling in the Pedigree Chart

Completely fill in as many of the spaces on the pedigree chart as you can. If you know an individual's middle name, include it in the chart, rather than using the initial. Put nicknames in quotation marks after the given name. Use women's maiden names; if you don't know a woman's maiden name, use a blank parenthesis, ( ). When you are able to do so, indicate the town or township, county, and state for any geographic places you include. Write all dates with the day first, the month next, and then all four digits of the year.

Because a pedigree chart is an instant guide to your research, keep it up-to-date so that you can follow its clues. Add new information to the chart as you find it, and correct any inaccurate information as soon as you've uncovered errors.

Pedigree charts cannot stand alone as evidence of your ancestors. They're research tools—not the end result of your research. Consider pedigree charts as notes to yourself that show, at a glance, where you are in your pursuit of your family history.

Family Group Sheets

Each of your ancestors is considered a family member, first as a child and then as a mother or father. You collect information about the entire family—not just ancestors, but also collateral relatives—to learn more about your ancestors. Record the information on your whole family on a *family group sheet*.

DEFINITION

The **family group sheet** is a form that allows you to record information on a family unit. It is neither an ascendant chart, nor a descendant chart; rather, it's a form on which all the members of your family can be listed (not just your ancestors).

Keeping You Organized

Family group sheets are the foundation for organizing everything you learn about your ancestors. The layouts of family group sheets may vary, but the categories of information are basically the same.

The top section of the record is designated for recording the names of a husband and wife, and the following information about each:

- Birth dates and places

- Christening dates and places

- Marriage dates and places

- Death dates and places

- Burial dates and places

- Occupation, military, religion

- Names of their parents

- Names of other spouses

Immediately following this information, you can record the children who result from this marriage with their birth dates and places; their marriage dates and places and spouses' names; and their death dates and places. If your family is a large one, you may need to continue the list of children on a second page.

When you gather information, collect it on whole families (your ancestors and their siblings). Each person on your pedigree chart should appear in two family group sheets: once as a child with parents and siblings, and then as a mother or father with children.

TREE TIPS

Make a separate family group sheet for each marriage of an individual. If a woman was married three times, then you should have three family group sheets for her.

FAMILY GROUP SHEET   NO. ____

HUSBAND [Full name] *William Nelson Marvin*		**SOURCES: Brief listing.**
BORN *2 Jan 1844*	AT *Harners, Onondaga Co., N.Y.*	No.¹ *1900 census*
CHR.	AT	No.² *Death Certificate*
MAR. *1 Oct 1875*	AT *Joliet, Will Co., Ill.*	No. 3. *1880 Census*
DIED *27 Jun 1904*	AT	No. 4. *Obituary*
BURIED AT *Braceville-Gardner Cemetery, Grundy Co., Ill.*		No. 5. *marriage return*
FATHER *George Nelson Marvin*	MOTHER [Maiden Name] *Mayetta Hammond*	(Complete source citations on reverse)
OTHER WIVES		
RESIDENCES		RELIGION
OCCUPATION *Farmer*		MILITARY

[Use separate forms for each marriage]

WIFE [Full maiden name] *Diana Courell Drollinger*	
BORN *22 Feb 1857*	AT *Roanon,*
CHR.	AT
DIED *11 Jun 1946*	AT *Joliet, Will Co., Ill.*
BURIED AT *Braceville-Gardner Cemetery, Grundy Co., Ill.*	
FATHER *Mathias Drollinger*	MOTHER [Maiden Name] *Amelia Jane Lloyd*
OTHER HUSBANDS	

Sex	Children	Day-Month-Year	City/Town County State	REF. No.
F	1. *Myrtle Olive Marvin*	b. *12 Oct 1879*	at *Grundy Co., Ill.*	Ref. *3,1*
	Spouse: *Jack Redwood*	m.	at	Ref.
		d. *8 May 1941*	at *Evergreen Park, Cook Co., Ill.*	Ref. 2
F	2. *Emma Grace Marvin*	b. *21 Dec 1882*	at *Grundy Co., Ill.*	Ref. 1
	Spouse: *Don C. Ritchie*	m. *4 Apr 1905*	at *Joliet, Will Co., Ill.*	Ref. 5
		d. *30 Dec 1943*	at *Detroit, Wayne Co., Ill.*	Ref. 2
M	3. *Sylvanus Jay Marvin*	b. *6 Apr 1885*	at *Grundy Co., Ill.*	Ref. 1
	Spouse:	m.	at	Ref.
		d. *25 May 1947*	at *Joliet, Will Co., Ill.*	Ref. *2,4*
F	4. *Ida May Marvin*	b. *31 Jul 1888*	at *Grundy Co., Ill.*	Ref. 1
	Spouse: *Clyde Edgar Dunn*	m *24 May 1908*	at *Joliet, Will Co., Ill.*	Ref. 5
		d. *15 May 1963*	at *Bloomington, McLean Co., Ill.*	Ref. 2
F	5 *Etta Jane Marvin*	b. *20 Jun 1892*	at *Grundy Co., Ill.*	Ref. 6
	Spouse:	m.	at	Ref.
		d. *18 Oct 1893*	at *Grundy Co., Ill.*	Ref. 6
F	6. *Sarah Bernice Marvin*	b. *21 Jul 1895*	at *Grundy Co., Ill.*	Ref. 1
	Spouse: *Rodney McKell*	m. *11 Apr 1911*	at *Salt Lake City, Salt Lake Co., Utah*	Ref. 5
		d. *28 Jun 1981*	at *San Dimas, Los Angeles Co., Calif.*	Ref. 2
	7.	b.	at	Ref.
	Spouse:	m.	at	Ref.
		d.	at	Ref.
	8.	b.	at	Ref.
	Spouse:	m.	at	Ref.
		d.	at	Ref.
	9.	b.	at	Ref.
	Spouse:	m.	at	Ref.
		d.	at	Ref.
	10.	b.	at	Ref.
	Spouse:	m.	at	Ref.
		d.	at	Ref.

PREPARED BY	OTHER MAR. OF CHILDREN
Address	Use reverse for additional marriages
City and State	Zip
Date Prepared:	

b.=born   m.=married   d.=death   ch.=christening   List references at top and use ref. numbers on items. Use reverse if necessary.

Here is a sample family group sheet.

Family Group Sheets and Sources

Each fact on the family group sheet must be linked to a full source citation. If all the information came from the same source, the family group sheet notation might read as follows:

> All information on this family from 1860 U.S. census, population schedule, Noble County, Indiana, Perry Township, Ligonier, page 105, dwelling 84, family 84.

As your research progresses, you will start to gather your information on the family from several sources. The same information may be in two or more documents. For example, you may have obtained the birth date of Mary Smith—2 March 1846—from both a family Bible and her tombstone. When you enter her birth date on the family sheet, list those two sources, and key them to her birth date. (When you enter a citation, be sure that you give sufficient information to properly identify it. See Chapter 20 for more on citations.)

Each fact on the family group sheet should be documented by a specific source citation. It is not enough to just add a list of sources to the group sheet with no indication as to which facts they refer. Those using the information later will need to know where you obtained the information on each individual fact: every birth, every birth place, every marriage, and so on. Use both sides of the paper if necessary.

Descendant Charts

You will likely use descendant charts later in your research. These charts start with an individual and list that person's descendants. Because descendant charts begin with a *progenitor*, you must do some research to find the progenitors in your lines. Descendant charts include all the descendants of the progenitor, or as many as can be identified.

DEFINITION

A **progenitor** is an ancestor in a direct line. When genealogists refer to a progenitor, they usually mean the earliest proven person in a line. If the earliest proven person in your paternal line is your grandfather, then he is the progenitor for that line. If you later prove who your great-grandfather is, then he becomes the progenitor for your paternal line. Each of your lines has a progenitor.

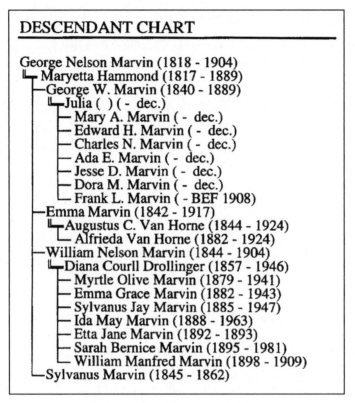

Here's a text-based descendant chart.

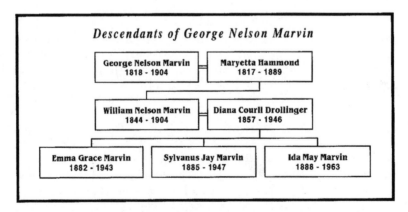

Here's a graphic descendant chart.

Drop Chart

A variation of the descendant chart is the *drop chart*. Sometimes called a box chart, the drop chart is a clear representation of one line of descent.

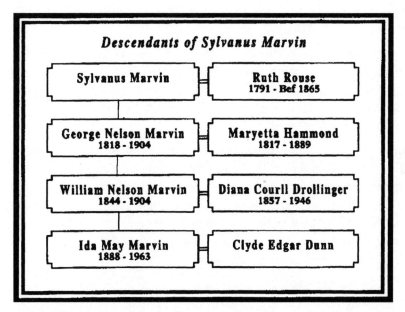

Descendants of Sylvanus Marvin

Sylvanus Marvin	Ruth Rouse 1791 - Bef 1865
George Nelson Marvin 1818 - 1904	Maryetta Hammond 1817 - 1889
William Nelson Marvin 1844 - 1904	Diana Courll Drollinger 1857 - 1946
Ida May Marvin 1888 - 1963	Clyde Edgar Dunn

This is an example of a direct descendant chart or drop chart.

DEFINITION

A **drop chart** connects two people, generation by generation, through their direct line. The researcher starts with the ancestor and then lists one child of the ancestor on the chart, let's say a son. Then, the researcher lists one of the son's children—the ancestor's grandchild—and so on down the line. A drop chart does not include all of the progenitor's descendants, only those in one direct line.

Notice that the last three figures use individuals from the pedigree chart and the family group sheet to illustrate the various ways you can treat your genealogy information. The first figure is a descendant chart for individual number 4 on the pedigree chart. The second figure is a graphic descendant chart for this individual, one of his sons, and some of his grandchildren. The third figure in the group is a drop chart showing the direct line connection from individual number 8 on the pedigree chart to his great-granddaughter, number 1 on the pedigree chart.

Using Computer Programs for Genealogy Record Keeping

Consider using a computer genealogy program to manage your genealogy information. These programs give you great flexibility in compiling and analyzing your research, as well as making it easier to share data with others.

After entering the data into the computer once, you can arrange it in many ways. All the programs use a family group sheet format to keep track of your information, and they all print pedigree charts. The differences in genealogy programs vary in the amount of information they allow you to enter and what you can do with that information once it is entered.

The programs range from very simple ones, which allow you to create only an organized collection of names and have the capacity to print basic forms, to very sophisticated programs, which produce customized reports and allow for extensive research notes, footnotes, and bibliographies to help you produce complete family histories. Many programs incorporate multimedia and provide tools to connect you to the Internet and to put your genealogy on your own webpage.

PEDIGREE PITFALLS

Avoid electronic genealogy programs that do not provide adequate space for citing sources. Look for programs that advertise full documentation of complete sources for each piece of your information.

Computer-Generated Reports

Computerized genealogy programs are especially helpful for preparing charts, both ascendant and descendant. The program knows from the information you enter just what names and dates it needs to gather from your lineage-linked database to compose the chart you select. Using a computer program makes it easy to print new, correctly numbered charts as you add newly found information.

Putting together a complete descendant chart by hand is tedious work. You must scour your files for all descendants and group them by generations. This is a task best suited to computer programs. In moments, a computer program can search through all the material you've entered and find everyone connected to any person you designate as the starting individual. It can then create a descendant chart based on your preferences.

Having all your data in a computer program makes it easier to write research reports and to share information with others working on your lines. A good genealogy program allows you to enter everything you find. It provides space to write evaluations of your data and sources, and space for remarks to yourself and for recording research tasks, such as "Check the 1850 census records of Indiana to find sibling's family." Once you've added notes regarding research tasks, see if your software includes an option to print your to-do list.

LINEAGE LESSONS

Most commercial programs have a built-in conversion program, so that if you switch to another genealogy program, you do not have to reenter your data. Look for Genealogical Data Communications (GEDCOM-compatible) programs. However, these conversions are not fail-safe; you will have to check your data carefully for errors that may have occurred during the conversion process.

Computer Searching: What Do You Want to Know?

With a good computer program you can search your data for nearly anything. You can look for all the individuals in your database born before 1900 in Grundy County, Illinois, obtain a list of every person in your database who served in the military, and find the average age of marriage for all the females in your database. Just creating a list of all the marriages in your database may be helpful to you in your research.

LINEAGE LESSONS

Most programs let you view your materials in different ways, making it easy for you to connect people. You can search for patterns or interesting statistics. How many men died of heart disease before age 60? Find all the women in your database for whom you need maiden names. How many relatives share your birth date? Remember, these searches depend on the data you have entered. If you have not entered information about military service, then the computer cannot give you statistics on how many men in your database served in the military.

You can determine your relationship to everyone in your database. How are you related to all the individuals you've researched so far? (If you collect records on several hundred people, you will inevitably have difficulty figuring this out without using genealogy software.)

Multimedia

Some sophisticated software programs are able to store photographs of your ancestors. All those charming photographs you have gathered—Aunt Lizzie on her first bicycle or Grandma tending her garden—can be stored on the computer and used to enhance your family histories. Those programs also allow audio and documents such as a PDF or a Word file to be attached to the person's other data.

Distant family members may not have seen the photos you can now incorporate into your genealogy. Pictures of the rude log cabin or the primitive sod house add immeasurably to descendants' understanding of what life was like in the early days of the United States. Imagine your children's glee as they look at the very bushy eyebrows on third Great-Grandmother Harriet and discover the origin of their own ample eyebrows.

Some programs allow you to incorporate audio clips into the material you gather. Grandpa recollecting his capture at the Battle of the Bulge or your cousins in Germany explaining the original pronunciation of your name add color to your family history.

Computer programs today are not usually bundled with CD-ROMs of resources, indexes, or family trees, as they were at one time. However, those CD-ROMs are still used by some individuals working on their family history. Be cautious about accepting the information on them at face value. Many of the lineages on the CD-ROMs that were once bundled with computer programs don't include source citations. Other materials in those bundled products, such as the Social Security Death Index, are now out-of-date and some resource materials were incomplete or had serious abstracting flaws. Though they may provide clues to further your research, you'll need to research and document the information you obtain from these CD-ROMs to be sure it is correct.

Continuing Development

There are constant technological advances that can and have had an impact on genealogy. The proliferation of personal data assistants led to the development of genealogy programs for mobile units. You can carry all your genealogy information in your cell phone. Utility programs (small software applications often called add-ons) work with major genealogy programs to enhance the products. These add-ons may facilitate printing fancy charts or preparing slide shows, or they may provide data analysis and make suggestions for specific resources to help you find the missing pieces. Many programs will also automatically generate pages that you can upload to the web.

Finding the Right Computer Program

If you are interested in buying a computer genealogy program, Cyndi's List (cyndislist.com) provides links to the many programs' sites. On the homepage, click on Categories and then select Software & Computers to access the list. You can also visit http://genealogy-software-review.toptenreviews.com for an exhaustive list of genealogy programs and reviews of each. Most programs have demonstration versions you can take for a "test drive" by downloading a demo from the site. Trying out the programs will make you more knowledgeable about the programs as you compare their features by reading reviews or by reading the message boards devoted to computer software (for example, rootsweb.com). Many local genealogy societies have computer interest groups that meet regularly for demonstrations and discussions of various programs. Computer programs change constantly, so make your choice based on the latest information you can find.

If you have an iPod, iPad, Android, or other mobile device, apps may be available that let you search vital records or other helpful information. Check the appropriate online store for your device for possibilities.

If you have any hesitancy on what program may be suitable for you, check the program's website to see if they offer a free trial download before you purchase.

The Least You Need to Know

- You can use ascendant and descendant charts to assist you in your record keeping.
- The family group sheet is a valuable chart for recording your whole family.
- Charts and forms are tools to use in your research; they are not your finished genealogy.
- Computer programs enable you to generate reports in many ways.
- Advancing technology continues to allow for the development of exciting aids for genealogical research and information sharing.

Finding the Trail

By this time, you are likely searching for the relatives you never knew. This part will make a detective out of you, guiding you in techniques to find your kin.

You'll start with the World Wide Web and be so staggered by all the wonderful things there, you'll think "everything is online." But it isn't so, you'll discover as you read on. So you won't be led astray, you will learn the variations in the names you will be searching. Then, you'll head for the library and find a whole world you didn't know existed. You'll learn how to use offline sources to supplement what you found online.

You'll even take another look at your family's hometown, seeing it from a new perspective. Now the town is a hunting ground—a place where there are potential clues for you to follow. All you have to do is find them!

The Internet: Online Any Time

In This Chapter

- Researching online 24/7
- Finding what you need
- Keeping track of where you've been
- Figuring out where to start your search
- Utilizing the wide scope of information

Technology's impact on genealogy cannot be underestimated. New resources become available hourly and all of them can be perused at your leisure. You're not bound by staffing constraints or budget reductions that close libraries and archives during the hours you are free. The Internet never sleeps—and if you're not careful, you won't either. That's the siren call of remote research.

In this chapter, you will learn all you need to know on how to access and understand online resources. Other chapters will add hundreds of website addresses you can use to conduct and broaden your search.

So, Everything Is Now Online?

First, let's dispel the myth that "everything is now online and it's free." In spite of the addictive nature of the Internet's instant gratification, you cannot complete all your genealogical research online. Genealogy information and evidence is culled from many disciplines and countless repositories, not all of which can afford to place their treasures online. Over 90 percent of genealogical manuscripts, records, documents, photographs, and other sources are not available online, so it wouldn't be realistic to expect everything about your ancestors to be accessible online.

The Internet's limits does not diminish the thrill of finding a scanned image of your great-grandfather's passenger list available online, seeing that he had brothers you knew nothing about, and doing it all in your pajamas at 3 A.M.! Still, take care when looking at online sources.

Online Information Can Be Wrong

Anyone can post family information on the Internet. There's no one policing the Internet to verify data or claimed relationships. You don't have to go far to encounter bad genealogical trees with erroneous information. Use these as clues to further information and resolve to publish your own correct tree (attached with sources of where you found the information) in order to show your corrections. When providing your online information to others be sure to suppress details about living people to ensure their privacy.

TREE TIPS

Internet information is not permanent, nor is it always up-to-date. When you find relevant information, capture it as soon as you see it. The data may not be there the next time you view the website, or the site host may have changed or moved what you needed. Recording the exact URL may help you find it in the future.

When Private Information Isn't Private

Genealogists love to share family information on the Internet. You may encounter your living relatives (or yourself!) in someone else's tree with complete birth date and children's names. If this bothers you, write to the author of the tree and request that they remove the information. Keep privacy in mind when you list your family information on the Internet, and seek permission from living relatives before putting their information online. Some genealogical programs will allow a substitution of "living" for names and block vital dates, thus effectively cloaking details on living people.

Online Information May Have a Price

While most genealogy information is free, some is available only by subscription or by paying fees to retrieve further information or documents. There is no charge to search the Social Security Death Index on most websites, but if you want a copy of a complete Social Security application (Form SS-5), you will have to pay a fee to the Social Security Administration.

The same information may appear on both free and paid subscription websites. Images of all available U.S. federal census records (1790–1940 population schedules) are accessible online through various websites. Understanding the extent of information on each site may help you decide which website best serves your needs. Not all subscription services are created equal. Their index and search capabilities may differ, so some may be better suited than others in helping you find your ancestors' names. Most subscription sites have a free trial period.

TREE TIPS

Try your census records search in several free and subscription sites to see index differences. Subscription sites may be available from your local library.

Traditional Methodology Still Applies

Even when you find online information, whether it's free or fee-based, you must analyze and evaluate it before incorporating the data into your family history. Do not integrate anyone else's work into your own without first checking it thoroughly. Use it as clues only. You don't want to inherit someone else's mistakes. The same basic research techniques you use to pursue family history in brick-and-mortar repositories must be applied to online investigation. Be systematic, working from the known to the unknown and from the present into the past one generation at a time. Keep a record of where you've been and what you found at each one. Cite sources completely, and evaluate your findings as you work.

If you download an online GEDCOM, create a new file in your computer. This way you can save the data with a note to yourself "to be checked" rather than incorporating the unproven data in a GEDCOM into your own work. And by keeping it separate, you can better judge what research you need to do to verify the information.

Okay, What Kind of Stuff Is Out There?

Genealogical information can come in many forms; each has its advantages and disadvantages. Scanned images of records allow you to see the information as it appeared in its original form. Databases, transcribed records, abstracts, and indexes all involve human interpretation of the originals, meaning they all have the potential to include some human error.

Nothing Like an Original Source

Scanned images of records are as close to holding the original as the online experience allows. You can see the handwriting and perhaps even your ancestor's signature. These records allow you to analyze them just as they were originally written. Being familiar with the family, you will see much more while examining the source than a stranger and be able to make connections. You may be able to bring down some significant genealogical brick walls when you view the original record.

Derivative Sources Drive Our Research

Typed transcripts of originals and abstracts of selected information may be thrilling to find, but remember: these documents were produced because someone looked at the original (probably handwritten) record and interpreted what they saw. In creating transcripts and abstracts, mistakes and typos can and do creep in. Treat these sources as clues until you can verify the information.

TREE TIPS

It's always a good idea to find the original record from which databases or abstracted information were derived.

Pointing the Way

Online library card catalogs and indexes to information each point the way to other information. Library catalogs may even point to scanned images of books. Indexes are meant to direct you to the original record. When looking at the Social Security Death Index or the California Death Index, verify the database information by sending for and viewing the original record.

Your Family Tree Is Online—You're Done!

You've found an online tree, but the thrill of its discovery is squelched by confusion and anger when you see it lists your great-grandmother married to her own brother or son. Or perhaps it states she gave birth to her daughter at age 8. These errors seem to take on a life of their own as they get passed from person to person. A large number of people using the erroneous information does not outweigh a single good factual source. Quality is better than quantity.

How do you correct such mistakes? You can try to bring it to the tree owner's attention, but your results may vary. They may not know how to correct their online tree, or worse, believe what they have is true. Another debunking technique is to list or post your correct tree with its sources so that no one can question how you came to know the correct information. Anyone who encounters both trees will see that yours has a better foundation.

PEDIGREE PITFALLS

Always verify the connections in online family trees with follow-up research. Anyone can post information online. No gateway guards check for sources or the accuracy of evidence interpretations. No authority certifies that posted materials are free of plagiarism and copyright violations. Be cautious about any online family tree and verify all its information with follow-up research before incorporating it into your own family history.

Online How-To Lessons

Aside from actual data and information leading to concrete data, there are instructional materials in the form of articles, *blogs*, e-zines, webinars, podcasts, videos, and online courses. Whenever you want to learn more on a topic or technique, consider these educational formats that may suit your available time and desired knowledge depth.

DEFINITION

A **blog** is a short article posted on a website and on which readers can comment. A dialog about the article may ensue. A blog writer may have a general focus, or he or she may write about a variety of topics. Some treat their blog as their website.

Online articles, blogs, and e-zines have various authors who make their material available through *RSS feeds* or by email subscription. They may have a specific focus, such as genealogical education, midwestern roots, or their family's history. Many people blog their research notes, which not only allows them to organize their research in one place, but also allows comments from anyone with knowledge of the family or who has research suggestions. With many hundreds of genealogy bloggers on the Internet, you may want to select a few interesting blogs to follow. Eastman's Online Genealogy Newsletter (eogn.com) is a popular place that brings together daily and weekly genealogy news items. Some blogs are listed at www.genebloggers.com.

Webinars allow a speaker to give a live lecture while the online audience views the presentation slides on their computer screen and hears the speaker's voice through their computer's speakers. Whether free or for a modest fee, you must sign up ahead of time for instructions on how to access the webinar at its appointed hour. Some lecturers record their live broadcasts and sell them. Once you confirm you want to attend, you'll likely receive an email that lets you add the webinar to your Outlook calendar. You will also get a link where you can read about video and/or audio requirements on your own system.

Some genealogy software makers, such as Legacy Family Tree and Roots Magic, offer webinars.

Podcasts (voice only) and videos are pre-recorded and you can play them at any time. FamilySearch (familysearch.org) has many instructional videos that you can also use for society programs. FamilySearch includes a video of the speaker's lecture along with the presentation slides; many also have a handout. Check iTunes for genealogy podcasts. Most podcasts are free.

Now How Do I Get to All That Good Stuff?

Now that you know the types of genealogical information out there, you need to identify the most efficient way to find what you are seeking. The chief search formats on the Internet are *directories* and *search engines*, and both lead you to webpages.

Some genealogy search engines act as the front end for the gold mine of data they index. Search boxes may limit you to first name, last name, and little else, or they may allow you to search by occupation or ship name. Massaging data through queries on these smart front-end engines, such as www.stevemorse.org, may help circumvent indexing typos and misunderstandings. Stevemorse.org allows you to search both free and subscription databases for immigration and census information. His "One-Step" site also is a valuable resource for those working with census enumeration districts, calendar conversions, language translations, and Jewish genealogy.

Genealogical Transporting Through Portals

If anyone can put up a website, how do you know which ones are the genealogically relevant ones? Categorized directories of links allow you to find information under various subjects, such as forms and charts, gazetteers, identifying females, and adoption research.

One of the oldest directories of genealogical data is Cyndi's List (www.cyndislist. com), which has over 300,000 links organized into almost 200 categories. Much like herding cats, Cyndi Howells conducts constant link updates that make the site a must-use for any project. Users of this website can suggest links or report broken links (that is, links that no longer lead to a website).

Try searching for detailed information by using search engines. Whereas directories provide you with links by searching for a subject, search engines allow keyword searches based on the terms you submit. The more specific your search request, the more relevant hits you will receive.

Aside from Internet browser searches, a general search of mocavo.com will garner results from free genealogical sites such as findagrave.com, archive.org, usgwarchives.org, and rootsweb.com. This is a good start to any online survey.

Asking Search Engines Nicely

Knowing how to speak the language of search engines will greatly increase the likelihood of a successful search. The following shorthand notations for creating a query to search engines will help you get finite results.

Double quotes: Putting double quotes around search terms, say *"Silas Harnden"*, tells the search engine those two words must appear next to each other. If a middle initial is included in the text on a website, chances are that occurrence won't appear in the results. (Be sure to try variant name spellings such as William Johnson, Wm. Jonson, etc. when formulating your query.)

Plus sign (+): Placing a plus sign in front of a search term means it must be present in the result. The two keywords *+Silas* and *+Harnden* with the plus sign preceding the terms, indicates that both results must be present somewhere on the same page. This is a good approach to use if there is some variance to the order in which a name appears on a site.

Minus sign (-): Placing a minus sign in front of a search term means the search engine must exclude results that include that term. The two keywords *+Harnden* and *–Michigan*, with the plus sign in front of *Harnden* and the minus sign in front of *Michigan*, indicates that the search results must include *all the Harndens* except for pages that also include the word *Michigan* on them.

Some search engines, such as Alta Vista (altavista.com), use *AND* when both terms need to appear on the page, *OR* when only one of the terms needs to appear, and *NOT* when pages that include a certain term that must be excluded. They also may use *NEAR* to indicate that two terms must be close to each other on the page. The latter is handy for name searches in which you don't know exactly how the name will appear.

LINEAGE LESSONS

Search engine keyword searches allow you to narrow a search. If you were looking for information on Wyoming genealogy, you could go to cyndislist.com or usgenweb.com, select Wyoming, and peruse the topics listed. If you want specific information about ranchers in Jackson Hole keeping their elk herd alive through the terrible winter of 1909–1910, you'd have more success using a search engine, such as Google (google.com). You may enter this key phrase: *winter 1909 Wyoming ranchers "elk herd"*.

Remember, there is no overall index for the World Wide Web, and no single search engine covers everything available. Additionally, Internet content changes constantly. Because a search engine is software that collects webpages and places them in a huge database, you are searching only that database, not the entire Internet. Search for the same thing on several search engines and repeat these searches over time. You'll likely get different results.

PEDIGREE PITFALLS

Although most search engines understand the plus and minus sign, you may get different results if you use *AND*, *OR*, and *NEAR*. Test it for yourself by using the same keywords with the different notations. Note how many hits return for each. *+Silas +Harnden +Maine* may return different results than *Silas AND Harnden NEAR Maine*.

A Simple Practice Exercise

Go to www.google.com, type your name into its search box, and review the results. Chances are you'll find someone who shares your name. To view the entries further, click on the one you wish to examine. It may surprise you to know that many individuals have the same name you do and that some of their personal information is close to your own.

This has further implications for family historians than just ensuring the correct name in the search engine results. The search engine may find an individual in an online family tree that seems to fit with the information you have. But does it really? Carefully weigh all the evidence before you claim a person as your own.

LINEAGE LESSONS

Search engines have help files and advanced search tips with examples. Take time to read them and try any tutorials that are offered. The more you learn about search techniques, the more successful you will likely be in your Internet search. By narrowing your searches, you are less likely to turn up thousands of irrelevant results. Cyndi's List (www.cyndislist.com/search-engines) includes links to articles and tutorials to help you understand how to build a search to return productive results.

Diving Into the Data Pool

General searches may lead you to the homepages of interesting websites that ask for data particulars in their own search boxes. They may ask for as little as a first and last name or as much as a birth date, place, occupation, immigration year, ship name, and more. You don't have to fill in all of the boxes. After all, some of what they're asking for includes what you're trying to find! By putting in too much information, you also may block some potentially valuable data hits.

Don't feel obligated to always search using surnames. Diving into the data pool with just a first name and a birth year and place may turn up a mangled surname index entry. Or, by entering no names for your ancestor's ship's list and only the ship's name and a town of origin, you may discover relatives and other associates of your ancestor who came with him.

Into the Wilds of Searching

Searching names using wildcards allows you to retrieve entries that contain a certain set and combination of letters. Ancestry (ancestry.com) allows wildcards as long as you provide at least three letters. Both *mar** and **gret* return Margaret and other variants. (The * represents the wildcard.) At stevemorse.org other search options can be chosen by utilizing the search form shown at the site. It provides such choices as "starts with or is" and "contains" certain letters, as well as various other options. You do not use the quotes when entering your search terms; just pick the option you desire on the form provided.

TREE TIPS

When a deep-diving search on one website fails to bring your ancestor to the surface, look to other sites that index the same records.

The indexes to record databases were created at different times by different people. Checking as many different indexes as you can will help get around cases where the index entries were mistyped or misinterpreted on one index but not the others. My ancestor Margaret Harnden was indexed as Mardell Hamden in one subscription database and as Marian Hamden in another for the same census record. How did I find her? She was living with her daughter and son-in-law who were correctly indexed. Looking for the extended family is another technique that's often successful.

Flexibility and creative searching are the keys to success. It is not that the data does not exist but that a problem may exist within the index that's preventing your short-cut searching from being successful. When the index search fails us, going line by line through a census county or several ship manifests may turn up your ancestor and other surprises. This technique is most successful when you know enough about your ancestor to narrow the search and recognize him or her in the records, even with spelling variations, ink blotches, torn pages, and incomplete data.

Recognizing Your Ancestor

Successful searches depend on using search terms that return relevant results. Sometimes the problem is not the search engine or database but the researcher who does not know enough about the research subjects to recognize them when they appear. Understanding that spelling, handwriting, and typos can obscure identities is part of the battle, and so is understanding what nicknames your ancestor may have used or what common ones were in use at the time. Polly and Mary, Sasha and Alexander, Nancy and Ann, are common variants.

Clerks who recorded the original information may have spelled phonetically. "Otto" may have been recorded as "Auto." Clerks also took shortcuts and recorded only initials. Some clerks misunderstood instructions to record last names first or may have recorded whole names backward so that the middle names were recorded as first names.

All of these variants come into play when you creatively query databases to obtain relevant data. A clever researcher will realize that his ancestor may never have spelled his name a certain way but that a clerk may have mistakenly indexed his name under an incorrect spelling. The results are only as good as the data you give it, so being aware of potential variants will help you to consider alternate searches and to recognize when the search results are the ones sought.

Perusing Printed Papers

Some of your searches may lead you to scanned images of printed books and newspapers. Typically, these were indexed using a computer program called *Optical Character Recognition (OCR)*. OCR's process introduces a new set of indexing problems, especially if the original document has stray marks, is crinkled or torn, or has blurry text. Computers can recognize letter patterns, but they can just as easily miss visual clues and assign the pattern to an incorrect letter. When searching materials that have gone through the OCR process, it is best to use various keywords to glean all the material on the subject.

DEFINITION

Scanned images are nothing but pictures of original documents, so you cannot simply run a search to find the words on their pages. Computers use **Optical Character Recognition (OCR)** to translate the content of these images into text. After going through the OCR process, you can then search the text in the same way you would search a Microsoft Word document. Unfortunately, the OCR process isn't perfect and may generate errors.

Dip Into the Layers

Have you ever gone to a party and encountered a layered dip? If you only scooped a bit from the top, you missed the delicious things below the first layer. The same is true of some websites. The best genealogical information may be several levels deep on the website, and the only way to find what you need is to explore beneath the surface. Hover your mouse over the graphics. Does the cursor change to a pointing hand? Click on the graphic to see what lies beneath it. It may only be a paragraph explaining a photograph; then again, it may be a detailed recitation of an historical event with links to other material on this or other sites. You can't predict what you'll find on a website.

Lost in Space

While randomly meandering in space, following links from one exciting site to another, is fun, it isn't very productive—especially if you don't record where you went or what you found. Intellectual curiosity is a plus, but with the temptation of a global library as close as your mouse, it is easy to spin out of control like an unbalanced washing machine and come to a distinct thud when you become totally unbalanced. The key to efficiency is planning and organizing and keeping a detailed research log.

A Global Positioning System (GPS) is not going to help you, and you may want to retrace your steps. Some Internet browsers have the ability to capture the history of the sites you visit, but don't depend on them in your research. Keep a paper or electronic record of where you've visited during your online search.

Whether you use a variation of the research logs mentioned in Chapter 3 or an electronic log, establish the habit of recording in them as a permanent part of your Internet experiences. You are more likely to record impulsive searches if you have a handy way to do it that requires little thought on your part.

Using an electronic log is expedient. Select the URL displayed in the browser, right-click on it with your mouse, and select Copy. Then go to your log, right-click where you want the URL to appear, and select Paste to add the link to your log. Not only is this approach convenient, it also reduces errors you may make if you are writing down a particularly long URL.

TREE TIPS

If you are familiar with your software's word processing, spreadsheet, or database programs, you can use any of them to create an electronic log.

Your log, whether on paper or electronic, should include the date, the URL, the name of the website, the search parameters, and the results. You can add columns for comments and surnames, but too many requirements may discourage you from keeping records consistently.

Bookmarks or Favorites

Use your research log to keep tabs on all the websites you visit in every research session. Use the bookmarks feature (sometimes called "favorites") in your browser to keep track of the places you want to revisit. Bookmark sites you expect to visit repeatedly or ones you access so infrequently you'll forget their addresses.

To keep your favorites list from becoming unwieldy, catalog your collection in folders, organized by folder titles. You may catalog by states, counties, surnames, or other topics of your choosing. Whether for genealogy or general use, arranging your bookmarks in a folder hierarchy results in more efficiency than scrolling through a list of 200 bookmarks to find the one you want. You can also store your bookmarks online at free bookmarking sites such as Digg, StumbleUpon, del.icio.us, Google Bookmarks, Yahoo! Bookmarks, Furl, Mixx, and Mister Wong.

Set Your Sights on These Sites

Many websites are important to every genealogical researcher. I'll mention some of these throughout this book, but here are a few you can go to right away to get a taste of what's out there in cyberspace. Start with these and you will be engaged for hours.

The first three sites listed below depend almost entirely on volunteers to run the sites and gather, *digitize*, or transcribe the materials. You may even find that you have information to contribute to one of the pages.

DEFINITION

To **digitize** is to convert data to numerical form for use by a computer. Digital representations (such as documents, maps, and photographs) can be stored and manipulated electronically, and displayed seamlessly on your computer.

The USGenWebProject

Started as a volunteer project in 1996, this grassroots effort by a small group of dedicated genealogists quickly became an important resource for family history researchers. The emphasis is on the county level, and the ambitious objective is to have an active website for every county in the United States. The sites vary from those that showcase extensive records and links to those that display merely a paragraph or two on the county.

Go to usgenweb.org and select a state. Explore what's there, then select a county to examine. You may find an index to early land records, cemetery transcriptions, or an Old Settlers list. There may be digitized maps and county directories. Perhaps you will find Civil War rosters or indexes to probate records. Most counties have a list of people who volunteer to do "look-ups" in their own reference works or in material at the library or courthouse. Often there will be a surname list and a query list. Many USGenWeb county sites have county-only search engines. You can also search back through years of submitted data by searching the USGenWeb Archives at http://usgwarchives.net.

RootsWeb

Perhaps the most well-known resources on RootsWeb (www.rootsweb.ancestry.com) are the RootsWeb Surname List (RSL) and the thousands of *mailing lists* and *forums* on surnames, localities, ethnic groups, software, and hundreds of other topics of interest to family history researchers. The RSL is a surname registry for researchers looking for more information about the described families. The registry consists of over a million names, each with a shorthand description of their locations and time periods and an email contact for the submitter. Another RootsWeb project is World Connect (http://wc.rootsweb.ancestry.com), a user-submitted collection of family trees.

DEFINITION

A **mailing list** allows those with a common interest to subscribe to the list and receive emails of all messages posted. RootsWeb.com hosts over 32,000 surname, geographic, and ethnic mailing lists. These are often archived for later searching or browsing. On a mailing list or **forum,** you can discuss messages of common interest, and you have the option to have the forum postings sent to you via email if you subscribe. Chapter 6 contains more information on mailing lists and forums.

FindaGrave

FindaGrave (www.findagrave.com), a volunteer effort to document gravestones from all over the world, is a growing database. Some cemeteries have readings of all stones or just a few, and some have photographs of all or some of the stones. Some cemetery cites include obituaries that accompany the stones. Visit this site periodically because it receives frequent additions. If you join FindaGrave (for free), you can also request that a volunteer take a photo of a tombstone for you. Volunteer photographers list the areas they're willing to travel. Submit your request, and if someone has volunteered in your designated area, they will receive an email notification of your request.

FamilySearch

This site at www.familysearch.org is sponsored by The Church of Jesus Christ of Latter-Day Saints. Headquartered in Salt Lake City, Utah, the church collects information to help members identify their ancestors so they can perform religious ordinances for those ancestors. They have been microfilming the world's records since the advent of that technology and readily share that information with the public. No subscriptions or fees are assessed to use this site. The millions of microfilms available at the Family History Library and through its local FamilySearch Centers are currently being digitized and indexed by volunteers at the time of this writing. It will take time for these records to become available on their website. Contributed pedigree charts, instructional videos, and other information are also available. See Chapter 8 for more about what you can find in their vast library materials, use of the catalog, and how to order microfilm.

Subscription Services

Numerous websites require that you subscribe before you can access information. Among them are Ancestry (ancestry.com), Genealogy (genealogy.com), Fold3 (fold3.com), HeritageQuest Online (heritagequestonline.com), and Genealogy Bank (GenealogyBank.com). Among these sites' many offerings are images and some indexes of the 1790–1940 U.S. census population schedules. The rest of their content varies considerably, so the best way to learn about their resources is to visit the sites.

Ancestry.com and genealogy.com sell subscriptions and products to individuals, but they also provide message boards and educational materials free of charge (even if you're not a member). Fold3.com has a few free databases, but the majority are accessible by subscription only. The subscription allows you to post a note on a record and create profiles of your ancestors. Fold3.com is continually growing from digitized federal records, city directories, and other records. Both Fold3.com (formerly Footnote. com) and genealogybank.com have free trials.

HeritageQuest Online (heritagequestonline.com) differs from the others in that it is only available to subscribing libraries that then allow their patrons to access the information. Some public libraries require cardholders to go to the library for access; others allow patrons with library cards to access the information remotely from home.

Who Else Is Online?

Social networking has changed the way we communicate with those closest to us and those who are friends of friends. It is a way to share photographs of our descendants and ancestors. Many genealogical societies now have Facebook pages (facebook.com) and Twitter accounts (twitter.com). Twitter is a great way to connect with researchers outside the United States, particularly with the United Kingdom, Scotland, and Ireland. Many state historical and genealogical societies have Twitter accounts; Twitter provides a good way to keep up on events on both a regional and national basis. The societies simply place their resources on it, within easy reach of their members and anyone else who "likes" them. LinkedIn connects business colleagues around common interests. Discussion forums focus on various topics, such as opening vital records in closed records state and county resources. Those who wish to find living relatives often turn to social networking sites.

What Else Is Online?

The resources for genealogy research defy itemizing. They range from digitized original documents to personal webpages. Things as diverse as vital records and online classes are yours for the looking. Social networking, mailing lists, and message boards can put you in touch with distant cousins and resources.

The Internet is dynamic and ever-changing. Websites reorganize, move, disappear, and reappear with different addresses. Whether you research in a traditional library or in the digital world, healthy skepticism is important when evaluating your finds. In both places, you will likely encounter extensive family trees carefully researched with full source citations for every fact, as well as family trees built totally on speculation with no source citations. The careful family historian follows up on all the material presented before believing it.

The Least You Need to Know

- No family history is all online, nor is all that is online accurate.
- Carefully constructed searches yield more relevant information.
- Keep a record of every website you visit.
- Explore the links provided on web pages.
- Trustworthy family histories are built on verifiable data.

Kissin' Kin: How to Find Them

In This Chapter

- Tracking down long-lost relatives
- Using family associations and reunions to find relatives
- Using the Social Security Death Index
- Using the Internet to find relatives

In earlier chapters, you learned how to start your search by using your own relatives as resources. But what about relatives with whom you have lost touch? To find these folks, you are going to get a chance to do some digging. To do so, you'll need to learn about a few records. If you have a bit of detective in you, wake up that side of your psyche. Start looking for family using these resources, which we discuss more fully later in this chapter:

- Telephone and other directory listings
- Online forums and message boards
- Libraries
- Newspapers
- Queries (published inquiries about the family)
- Family associations
- Family reunions
- The Social Security Administration

Ma Bell Comes to the Rescue

Your local library's reference department may have a collection of telephone directories covering the whole United States or at least a few states. Better yet, use the various online directories as available at argali.com, switchboard.com, and others. List the relatives who may still be living, and spend an evening at the library or online. If you cannot find the specific person you're seeking, search for all those who share that person's surname in the area where you believe he or she lived. If the list is not too long, you can easily do this at the online sites by specifying the surname and the town or city. For common names, such as Smith or Jones, you need to narrow down the search results by looking for familiar family-given names.

Prepare a letter giving some details of your family and your purpose and send it to each person on your list. If the phone book does not provide the zip codes for the addresses, check the library while you are there for their zip code directory. You can also obtain the zip codes at the United States Postal Services website, USPS.gov. On the homepage, click on Find a Zip Code, enter the street address, click Submit, and the site will supply the nine-digit zip code.

Several commercial companies used to issue CD-ROMs of residential and business telephone numbers. Their databases varied, but they were compiled from a variety of sources such as utility companies, mail-order houses, voter registers, and telephone books. The margin of error was great; people move often, so the lists became quickly outdated. Nonetheless, if you are looking for relatives and you think they lived in a certain area 10 to 20 years ago, you may find them listed on those old CD-ROMs. Check your library for availability.

Your state or city library or historical society may have a collection of old telephone directories for selected areas. The series is rarely complete, but it may help. When the listing disappears from the directory, it may signify death or a move from the area, providing you with leads. In recent years, the absence of the names cannot be easily interpreted, because parties now so often choose to keep their contact information unlisted. Some online directories are not only from phone listings, but also from a variety of other sources, and therefore unlisted numbers may have slipped in.

After obtaining the contact information for a possible relative, resist the temptation to pick up the telephone. A letter can assure the person that you are related, and may elicit more information than a surprise call. If you don't get a response, follow the letter with a telephone call. Your letter will have introduced you and your purpose, setting the person more at ease.

The lack of a response to a letter does not necessarily indicate disinterest. Some hate to write or can't write. Your initial letter may, however, serve its purpose as an introduction and open the lines of communication.

If you have the person's email address, you can use that; but be sure to include sufficient details.

Consulting Your Local Library

The local library's holdings depend upon the size of the area. Although they may be too small to have books that directly pertain to genealogy (see Chapter 8), they will no doubt have a number of traditional reference books that will be useful. Check the reference section. Besides the telephone books already mentioned, look for other types of directories that may be useful to genealogists.

City Directories

Become acquainted with city directories, found in your library's reference section. You will likely use these extensively, in a variety of ways, as your research techniques develop.

Your efforts may be highly productive in tracing the earlier generations of your family using city directories. Early directories list the residents alphabetically and often include a cross-reference by street address. This provides important details: the neighbors. If the neighbors' families still live there, they may have known your family. If they were close friends, they may even have kept in touch and may provide you with current addresses.

Most cities have changed the format of their directories. They continue to provide a listing by street (called a householder's index), but they have eliminated the alphabetical name listing. Instead, they usually provide a cross-index by telephone number. Although some towns still retain the old-style alphabetical format, you will most likely lose the ability to search by your family's surname in most current editions.

Write to the library of the town or city where you believe your family lived. Ask them to search the current issue of the city directory or to photocopy the pages with the surname or street you are seeking. When you're asking for specific dates, give them no more than one or two years, as most librarians are very busy. Offer to pay the costs, and enclose a *SASE* or *LSASE*

DEFINITION

SASE is a self-addressed stamped envelope. **LSASE** is a long self-addressed stamped envelope. Due to the high cost of postage, many organizations (and individuals) won't answer inquiries if the inquiring individual hasn't provided a SASE.

To obtain the addresses of libraries to which you can write, contact the reference librarian at your local library and ask the librarian to check the American Library Directory. You can also go to the American Library Directory (americanlibrarydirectory.com) and register for free. On this website, you can search for libraries, though with the free registration you will only be able to see the name and address. (Paid subscribers can access the complete listings of the libraries, which include a listing of the library's holdings.) In using their search feature, you will get only libraries starting with a city's name if you enter a city on the first line that asks for the library or institution name. If instead you choose to enter the city in the "City" field further down on the search form, you will get all of the libraries in that city.

Elizabeth Petty Bentley's *The Genealogist's Address Book*, or Juliana Szucs Smith's *The Ancestry Family Historian's Address Book: Revised Second Edition*, can assist with addresses, too, although the libraries listed in those sources are mainly those with genealogical holdings. Many small-town local libraries are not included in their books.

Posting It

In determining ways in which you might find your lost relatives, consider the library bulletin board. The librarian may be willing to post a notice. Mention that your research is for genealogical research, and enclose separate notice for posting. Keep it brief, but give enough details to identify the family you are seeking.

Ask the librarian if the library maintains files of letters from people who have written the library for information, often referred to as "vertical files." A letter from your second cousin, also interested in genealogy, may be in that file waiting for you to find it!

PEDIGREE PITFALLS

Never send a notice for posting on a library bulletin board with writing on both sides; it cannot be properly displayed. Be sure it includes your name, mailing address, and, if you have one, your e-mail address. Your phone number is optional.

Sign In, Please

Visitor registers are popular in the genealogy section of the library. A bound volume (or loose leaf, or even 3" × 5" cards) is usually positioned prominently for visitors to enter their names, addresses, and the surnames of interest. The librarian may be willing to search it for the surnames you're seeking. In some areas, the genealogical society has even indexed such entries.

Take a few minutes to add your name if you visit any facility with a guest register, such as a courthouse, museum, library, or historical attraction. Your reward for doing so may be a letter or telephone call from a distant relative.

The Hometown News

To locate living relatives, consider subscribing to their hometown newspapers. You may be amazed at the leads it can provide, particularly if its circulation encompasses a small community. Watch for family names in the social news, school news, church news, and even the advertisements. All will be of interest and may lead to contact with relatives. Be sure to record in which newspaper you found a lead, and note the page number and column. Try to get a microprint of the page while you are at it.

Don't overlook obituaries, particularly the part about "the survivors include ...." It is common to include brothers, sisters, aunts, uncles, and children, and often their cities of residence. This will help immensely.

Also try writing a letter to the editor of the hometown newspaper. This may be especially successful in smaller communities. Make your letter brief, but show clearly that it is a family local to their area and that your interest is genealogical.

Dear Editor,

My great-uncle John Jackson died 15 September 1925 in your town. At that time his obituary stated that he left two sons, Joseph and George, and a daughter Mary Jefferson. I am interested in contacting them as we are compiling our family genealogy. Can some of your readers put me in touch with them?

Thank you for your help.

Sincerely,

[Your name and address here.]

Here's a sample letter to the editor inquiring about living relatives to the newspaper.

Don't overlook the use of the classified section. An advertisement under "Personals" may be seen by those who know your family. Give a few details so that the family can be quickly identified; ask to reach your relatives. Mention your interest is in genealogy.

Take advantage of the online forums and message boards, too, especially those that are surname oriented. Post your query in a short but clear posting and then wait and see what happens. Many have found long-lost relatives using this approach. Don't limit your search to surname forums—try also those that are locality oriented. If you know that the family lived in Monroe County, New York, then post to a forum for that New York area. It may yield excellent results for you.

Some of the better-known forums and message boards include www.genforum. genealogy.com, and ancestry.com, among others. See Chapter 5 for further discussion. Some of the forums cloak the email address in such a way that you can use it but spammers can't harvest the email addresses for their soliciting purposes.

After School: Alumni Records

Does someone remember that Cousin Harry was a graduate of the state college or that his brother attended Harvard? Check for a college alumni directory on your library's reference shelves or go to the college's website for the school's address. Address your letter to the attention of the Alumni Director. The school may have alumni records with addresses. In 1919, Harvard, for example, published the *Harvard Alumni Directory: A Catalogue of Former Students Now Living: Including Graduates and Non-Graduates, and the Holders of Honorary Degrees.* This directory is available at http://distantcousin.com/yearbooks/ma/harvard/alum1919.

In the Harvard directory, I found more than one page of Walker alumni living in 1919 all over the United States and even in Canada. There is an amazing number of directories available on the Internet—many such directories are out of copyright, so there are no restrictions for posting them online. Also seek literature on school reunions. A newsletter after the event may include your relative's new location.

At the website distantcousin.com/yearbooks, there is a search feature for many school yearbooks and directories, categorized by state. This site also includes many other types of records and photos. Well worth your visit!

PEDIGREE PITFALLS

Be sure to type the URL correctly for Distant Cousin: distantcousin.com. If you add an "s" to the end of cousin, you will find yourself at a website designed to lead you to many commercial sites instead of to the good items at distantcousin.com.

The Genealogical Societies

When you have determined the county in which your relatives lived, obtain the name of the county's genealogical society. There are a number of directories that may assist you. Here's a list of sources that list genealogical societies:

- Everton's *The Handy Book for Genealogists*
- Ancestry's *Red Book*
- Ancestry's *The Source*
- Elizabeth Petty Bentley's *The Genealogist's Address Book*
- Juliana Szucs Smith's *The Ancestry Family Historians Address Book*

Examine these sources for addresses of genealogical societies in areas all over the country.

Alternately, use the Internet. Go to cyndislist.com, click on Categories, and select Societies & Groups. Make use of the USGenWeb Project by going to usgenweb.org. This amazing volunteer project strives to provide websites for every county in the country. Each website offers a diversity of information and records. They usually also list the libraries and genealogical societies for the county.

Consider joining the genealogical society in the county in which your relatives lived. As a member of that society, you will have privileges. You may find an interested volunteer who will assist you. If the organization publishes a newsletter, ask them to publish a notice in their *query section*. One of your relatives may even be a member of the society. Many such groups solicit family charts from members. Request a search of those charts for the names of others working on your family names.

 DEFINITION

A **query section** in a genealogical publication is a specific section of the magazine set aside for submitted inquiries. Almost all periodicals have such sections, as do some newspapers that carry genealogy columns. They may limit the length or have other restrictions.

Periodicals Galore

Thousands of genealogical periodicals are published each year in the United States. In addition to placing queries in those in the county you are researching, place them in others that are published statewide or nationwide. Inquire about cost and whether they require membership or subscription to use that service. Make some choices depending upon your budget. Mention enough about the family in your query so that readers can identify it.

Seeking descendants of John Masters

b. 15 April 1858 in Monroe Co., NY, m. there in 1888 to Mary Adams. They moved about 1910 to Perry County, IL and had 3 ch., John b. 1912, Joseph b. 1915, and Agatha b. 1918 (who m. John Davis). Wish to contact descendants.

This is a sample query to submit for publication in genealogical newsletters.

A number of national newsletters are connected to a society and are part of the society's membership benefits. You can subscribe to most whether or not you are a member: The New England Historical and Genealogical Society's *New England Ancestors*, The National Genealogical Society's *NGS Magazine*, and others. Some national periodicals are subscription only—that is, they aren't associated with a society. Some subscription journals focus on articles but do accept a limited number of queries from members.

LINEAGE LESSONS

It is permissible to use standard abbreviations in your query. For example, "b." is used for born, "m." for married, and "ch." for children. These reduce space.

Families That Stay Together

A valuable resource in your effort to locate relatives is a family association dedicated to the particular family name you're researching. These associations operate under a variety of names, such as a society or clearinghouse. Their purpose is to collect information on the family, share it, and preserve it. They may publish a newsletter, maintain a database for researchers, or offer search services. They are eager to hear from descendants. They may be able to supply the addresses of others in your family. If they publish a magazine, they can insert a query for you.

There have been thousands of family associations, but many are short-lived. Others are so short-handed that they often don't have the resources to reply to inquiries. Do not let the lack of a reply from one group influence your decision to contact another. Many hundreds maintain extensive archives and are eager to hear from you. Their records are valuable and sometimes are the only source for a particular photograph, a family Bible, or an old letter.

Locating a Family Association

Currently, there aren't any directories attempting to completely list groups known as family associations. For many years, Everton's *Genealogical Helper* published an annual listing, and old issues of this defunct magazine may be of assistance. Also try cyndislist.com by going to Surnames, Family Associations, & Family Newsletters under Categories.

Of course, you may try simply typing the family name followed by family association into your Internet browser; for example, Rose family association. Running this search may turn up listings, but it may miss those groups that don't include family association as part of their name. Try also a search using the surname plus the words newsletter or society or clearing house.

Elizabeth Petty Bentley compiled a useful *Directory of Family Associations* that had been updated periodically, but there doesn't seem to be a current edition as of this writing. Look for past editions of this directory in your library.

TREE TIPS

If you have difficulty locating an association, go to a service such as genforum. genealogy.com, and then proceed to the message board or mailing list for the surname. Post a query asking if anyone knows of a family association for that name.

Be aware that there are other types of associations that can be helpful, such as descendants of a named person (Pocahontas, etc.). Or the association may be geographically oriented; all those who came from a certain foreign area or all of those who settled in a certain American location.

A Rose Is a Rose Is a Rose

When using the resources of a family association, it is important to understand its focus because they differ considerably. They may include descendants of an ancestor who isn't the immigrant, descendants of the immigrant, or descendants of the surname.

Those focused on the descendants of an ancestor who isn't the immigrant, for example, may include all the descendants of the great-grandfather and his wife who settled in Des Moines, Iowa. In this case, the association is probably named after that couple. Groups devoted to descendants of an immigrant usually carry the surname in their title and include all the male and female descendants of that immigrant. In the third type, the surname organization, anyone bearing the surname is traced. They may have no relationship to each other. Members do not have to bear the surname, but each trace someone carrying that surname.

TREE TIPS

Although most family associations do not require you to join for them to be of assistance, ask for a copy of their membership brochure. You may find it beneficial to be a member of the group, depending upon their goals and services.

Even surname organizations have differences. One may search all of the surnames in the United States, regardless of nationality. Another may focus on an ethnic group, such as only those of Scottish ancestry or only those of German ancestry.

Upon ascertaining the existence of a family organization, write or send an email to them. Be specific in your request. Identify the family briefly (include names, dates, locations, spouses, and children), and ask if they can put you in contact with members of that family. Inquire also about publishing queries in their magazine or on their website.

We'll Meet on the Fourth Sunday ...

Thousands of families hold family reunions each year, ranging from a picnic in the park to a major gathering in a large hotel. Watch for announcements in genealogical publications. No magazine includes all the reunions, only those for which they have received an announcement. Try to attend if possible. The contacts you make will surely evolve into wonderful friendships, and you will hear some of the many family stories you have yearned to know. The reunion may be small, organized by a local branch, or nationwide, involving hundreds of descendants. Some even plan overseas trips to the ancestral home.

Using the Social Security Administration to Find Your Kin

In 1935, President Franklin Delano Roosevelt signed the Social Security Act into law. It generated an enormous amount of paper records, some of which you can use to locate your relatives.

PEDIGREE PITFALLS

Although the online databases are titled "Social Security Death Index," the databases only include deaths that were reported in order to receive a benefit, and the listings are therefore incomplete. The database was started in the early 1960s.

The Death Index

One especially useful resource is the Social Security Death Index (SSDI), previously available on CD-ROM and now available online at www.ancestry.com, at www.genealogybank.com, and at other sites. It does have its limitations. The database does not include all those who died with a Social Security number, only those whose deaths were reported to the Social Security Administration (SSA). Usually, deaths were reported to the SSA when the family applied for a benefit upon the death of their relative. If the individual's death was not reported to the SSA, then his or her death is not listed in this database. Consider also that some websites that offer this index update more frequently than others.

The information contained in the SSDI usually includes the following:

- Social Security number
- Name (last and first)
- Birth date
- Death date
- Issuing location of the original SSA card (by code)
- Zip code of last known residence
- Zip code of recipient if a lump-sum payment was made

Locating an entry on the SSDI may help you find living relatives, too. For example, after finding an individual on the SSDI, you can follow up on his or her date of death by obtaining the death certificate. The death certificate may lead you to living relatives by providing the place of death and the name of the informant, who may be related. Then check a phone or online directory (see earlier in this chapter) to find the informant.

You can obtain a microprint of an individual's Social Security application, the SS-5 form, on microprint. The microprint will supply the birth date and place of the applicant and the name of his or her parents. Doing this doesn't help in finding living relatives because you're only able to obtain the applications of deceased persons. It can, however, be tremendously helpful in tracing the family to earlier times. The great advantage of this source is that the applicant personally supplied the name of his or her parents, making it a highly reliable source. (There could be errors in that information if the applicant was misinformed, but in most cases it would be considered to have accurate information.)

If you found the correct listing on the SSDI, then you need only to send the current fee (available on the Social Security Administration's website) and the Social Security (SS) number of the deceased. Go to http://ssa-custhelp.ssa.gov and then click on Search. Fill out the form with the deceased's name and anything else you may know. After you submit the form, the results should be immediately displayed online. For a small additional fee, you don't even need to supply the SS number to get the SS-5. If you found the SS number in some way other than on the SSDI and you can't find the name included in SSDI, then you will be required to submit proof of death. Mail any handwritten requests for SS-5s to this address:

> Social Security Administration
> Office of Earnings Operations, OEO FOIA
> 300 N. Greene St.
> P.O. Box 33022
> Baltimore, MD 21290-3022

LINEAGE LESSONS

The Social Security Administration produces a master list, and this list is available to commercial companies that prepare the death indexes. Because these companies update at different times, it may be worthwhile to search more than one SSDI, especially if the death occurred recently.

The Internet: Finding Cousin John

There is an explosion of activity on the Internet that can help the genealogist. Many forums are available on which users can post a notice of interest in certain families, and others can respond. Use these message boards to seek relatives. Some report astounding results, locating lost branches of their family whom they had never expected to find. Try http://lists.rootsweb.ancestry.com/cgi-bin/findlist.pl or www. genforum.genealogy.com. Each has an alphabetical index to the surname boards. Also consider posting a query on one of the county sites available through www. USGenWeb.org.

As always, be specific when posting queries. Simply asking for help in locating relatives of "The Taylor family who lived in central Ohio" or posting the subject heading "Help" is too vague. Such postings will likely be ignored or induce overwhelming responses asking for clarification. Inquire instead with precise details. Insert some idea of time frame and location. For example, if you post a notice with the subject heading "George Taylor (1748–1810) Fauquier Co., Va.; Jefferson Co., Tenn." everyone knows immediately which Taylor you are tracing.

Use of extensive abbreviations in posting notices on the Internet is acceptable, but be sure you don't sacrifice readability. You want everyone to understand.

TREE TIPS

For tips on preparing queries, go to www.cyndislist.com/queries.htm. Then go to "How To." There are several informative articles here. A well-constructed query can dramatically increase results.

Not the End of the Tale

As your search progresses, you will constantly discover new ways to locate living relatives. A letter written by a descendant may be part of the military pension file in the National Archives, and it may provide you with the location of family members. The state may maintain a statewide property index and allow citizens to search it, sometimes even online. The voter registers may be open to the public. The possibilities are endless. You will be thrilled when your scouting turns up your people. Every success will energize you and fill you with the satisfaction of following clues and finding leads. You're now on your way to the same addiction afflicting the rest of us!

The Least You Need to Know

- Your library and the Internet have telephone directories, city directories, and other items to help you search for your living relatives.
- Your ancestor's or relative's hometown newspaper may assist in the search.
- Use genealogical societies and family associations, which can help locate living relatives.
- Send queries to online message boards and mailing lists.
- The voluminous Social Security Administration records are a wonderful tool for research.

A Rose by Any Other Name ...

In This Chapter

- Surnames change over time
- Deliberate or accidental changes
- Spelling variations
- Handwriting complications
- Naming patterns

Surnames emerged to help distinguish one person from another. They were descriptive of relationships, such as Thomasson to indicate Thomas's son; of physical characteristics, such as Petit indicating small and Schoen indicating beauty; of place names such as Hill, Lea, Meadow; or of occupations such as Carpenter, Hunter, and Miller. Many of us are known by surnames that would surprise our immigrant ancestors. Surnames are not static through time, nor do they necessarily stay the same for an individual throughout a lifetime.

The spelling of names is phonetic, and the way the letters are arranged depends on the person writing the name. Most of us descend from people who were illiterate. Our ancestors couldn't write and didn't know how their names were spelled. They knew only how to pronounce their names as spoken by their families and neighbors.

Variations Aplenty

Outside influences, unusual events, naming patterns, spelling variations, and misinterpretations by transcribers determined the surnames of our ancestors. Before searching for documents, consider that your ancestors' names may be different from what you were led to believe. The same surnames may be spelled so many different ways that the name may be unrecognizable to someone trying to trace the line. For example, in the following figure, each of these surnames was used by at least 100 people at the time of the 1790 census.

Ministers, recording clerks, ship captains, and census takers wrote what they thought they heard. And what they heard was based on their knowledge and background. Language and spelling were not standardized, even among the educated, until more recent times. Many times a name was spelled several different ways in the same document. Standardization evolved slowly and varied in different areas, but was particularly noticeable in the nineteenth century.

Fitzgerald, Fichgerrel, Fitchgearald, Fitchgerrel, Fitsgarrel, Fitsgerald, Fitsgerel, Fitsgorrel, Fitsjarald, Fitsjerald, Fitts Gerald, Fitzarrell, Fitzgarald, Fitzgarrold, Fitzgearld, Fitzgeral, Fitz Gerald, Fitzgerrald, Fitz Gerrald, Fitzgerrel, Fitzgerrold, Fitzjairald, Fitzjarald, Fitzjerald

Rawlings, Raling, Rallins, Raulens, Raulings, Rawlins, Rollens, Rollin, Rolling, Rollings, Rollins

Sinclair, Saintclair, St. Clair, St. Clear, St. Clere, Senkler, Sinekler, Sinclar, Sinclare, Sinclares, Sinclear, Sincleer, Sincler, Sinclere, Sinclier, Singclair, Sinklar, Sinklear, Sinkler

---

These name variations appear in the first United States federal census, 1790.
Taken from *A Century of Population Growth.*

Handwriting Further Obscures the Names

Letter formation often leads to difficulties in reading and interpreting surnames. Flourishes and curlicues can render letters nearly indecipherable, particularly to the untrained eye. The letters I and J sometimes look very similar. A poorly written G can resemble an S. The letters R and K are often written similarly, and a capital B that is not closed at the bottom may be mistaken for an R. Other letters also can be confused, such as M and N, N and H, and V and U.

Indexers had difficulty interpreting letters, too, so the name you are looking for may be alphabetized under the wrong letter. If you don't find Seemon in the S section, look under the L section.

TREE TIPS

Sometimes looking at other documents or entries written in the same hand makes the letters easier to read and recognize. Comparing the names in the county as written by two different people, let's say the county clerk and the tax assessor, may clear up any handwriting-related confusion.

Immigration Changed Lives and Sometimes Names

Immigrants often changed their names by accident or by design. Some languages have letter combinations not found in English, or sounds signified by other letters in the English language. *F* sounded like *V*, so Freer was written Veer. *W* sounded like *V*, so Werner was listed as Verner. A guttural pronunciation made *G* sound like *K*. Letters were slurred and *H* dropped. *Sch* sounded like *Sh*, so the *c* disappeared.

You can get some idea of how names changed and how a name change might affect the immigrant by reading an excellent article by Marian L. Smith titled "American Names: Declaring Independence," at www.ilw.com/articles/2005,0808-smith.shtm. I also recommend reading "They Changed Our Name at Ellis Island" by Donna Przecha, at www.genealogy.com/88_donna.html.

The Wish to "Sound" American

Occasionally, immigrants shortened or changed their name, but more often, the children of these immigrants Anglicized their names to better assimilate: Petrasovich became Preston, Noblinski became Noble, Savitch became Savage, Madsen became Madison, and so on.

Sometimes feeling their names were "too different," immigrants chose to translate their names to the English equivalents. Blau and Bleu became Blue; Weiss, Blanc, and Bianco all became White. Occupational names were changed with Schmidts becoming Smiths and Küypers becoming Coopers.

If you are unfamiliar with the language of the immigrant you're researching, look for a dictionary that has both a foreign language and English. Perhaps the foreign language equivalent for the surname will be the one you need to further your research. Knowing that "Weaver" translates to "Weber" in German may be just the clue you need to find an earlier generation.

A wish to disassociate from "foreigners" made some families make the change. In Pennsylvania, Frank Nicotera felt the pressure of his Italian name. When he married a Collins, he registered at the union hall as Frank Collins. When his little brother was ready to go to work, Frank took him to the union hall where everyone assumed he was Nick Collins, not Angelo Nicotera. To this day, Frank's descendants go by the name Collins. The grandchildren are unaware that the family's surname was Nicotera only three generations ago.

The following Declaration of Intention is an example of the name changes that happened often. According to his birth certificate, the subject was born 20 October 1892, in Casal Velino, a province of Salerno, Italy, and his name was Florigio Cicerelli. He first immigrated to the United States in 1910 (using his name with that spelling), along with his father Saverio, on the *U.S.S. Berlin*. They arrived in New York City and headed west toward San Jose, California. Later returning to Italy, Florigio married Carmela Spagnuola in 1922, again using the same spelling. When Florigio returned to the United States in 1930 from his fourth and final trip to Italy, the Arrival Certificate in New York shows him as "Cicierelli." Three years later, when he filed to become a citizen, his name appears "Ciciarelli." In the meantime, he also had adopted the more American sounding "Frank," and was known for the rest of his life as Frank Ciciarelli. His children's surnames bore that spelling, too. Similar changes occurred in thousands of families.

Families make unpredictable decisions that cause research problems for genealogists. Letters are added or dropped. Perhaps the name had a prefix that was later removed: St. Martin became Martin and Van Hoorn became Horn. A generation later, someone may have chosen to reinstate the suffix. One branch of the family may have decided the name would be more aristocratic with a slight change. Even simple names undergo changes. Instances abound in which families add or remove the "e" from their names: Brown/Browne, Green/Greene, Germaine/Germain, Low/Lowe.

LINEAGE LESSONS

There are families in which brothers have different surnames even though they have the same father. One has decided to keep the old spelling, whereas the others have adopted a new spelling.

These variances sometimes lead to two branches of the same family having very different surnames. In one case, a divorce in New York City in the 1820s resulted in the wife resuming her maiden name and her younger children taking that name, while the adult children retained the father's name.

TRIPLICATE
To be given to
declarant)

No. 5135

UNITED STATES OF AMERICA

DECLARATION OF INTENTION
(Invalid for all purposes seven years after the date hereof)

State of California

County of Santa Clara

ss:

In the Superior Court

of California at San Jose, Calif.

I, Florigio Ciciarelli
now residing at Route 2, Box 5, Los Gatos, Santa Clara, California
occupation Barber, aged 40 years, do declare on oath that my personal description is:
sex Male, color White, complexion Medium, color of eyes Blue
color of hair Brown, height 5 feet 8 inches; weight 165 pounds; visible distinctive marks None
race Italian, North; nationality Italian
I was born in Casal Velino, Italy, on October 22, 1892
I am married. The name of my wife or husband is Carmela Ciciarelli
we were married on Oct. 2, 1923, at Casal Velino, Italy; she or he was
born at Casal Velino, Italy, on July 17, 1900, entered the United States
at New York, N. Y., on Jan. 16, 1930, for permanent residence therein, and now
resides at Los Gatos, California. I have two children, and the name, date and place of birth,
and place of residence of each of said children are as follows: Renato, born Dec. 7, 1929, at Naples,
Italy; and Angela, born Apr. 27, 1932, at Los Gatos, Calif.; both re-
siding at Los Gatos, California.

I have heretofore made a declaration of intention: Number ___, on Aug. 3, 1923
at San Francisco, California, in the U. S. District Court.
My last foreign residence was Casal Velino, Italy.
I emigrated to the United States of America from Naples, Italy
my lawful entry for permanent residence in the United States was at New York, N. Y.,
under the name of Florigio Ciciarelli, on August 23, 1910,
on the vessel Berlin

I will, before being admitted to citizenship, renounce forever all allegiance and fidelity to any foreign prince, potentate,
state, or sovereignty, and particularly, by name, to the prince, potentate, state, or sovereignty of which I may be at the time
of admission a citizen or subject; I am not an anarchist; I am not a polygamist nor a believer in the practice of polygamy; and
it is my intention in good faith to become a citizen of the United States of America and to reside permanently therein; and I
certify that the photograph affixed to the duplicate and triplicate hereof is a likeness of me: So HELP ME GOD.

Florigio Ciciarelli

Subscribed and sworn to before me in the office of the Clerk of said Court,
at San Jose, Calif. this 7th day of August
anno Domini 33. Certification No. 22-12153 from the Commis-
sioner of Naturalization showing the lawful entry of the declarant for permanent
residence on the date stated above, has been received by me. The photograph
affixed to the duplicate and triplicate hereof is a likeness of the declarant.

Henry A. Pfister

[SEAL]

Clerk of the Superior Court.

By _____, Deputy Clerk.

Form 2202-L-A
U. S. DEPARTMENT OF LABOR
NATURALIZATION SERVICE

This Declaration of Intention form was filed in 1933.

Talking to Yourself

You may need to be creative to find other spellings of the surnames you are tracing. Think of the many ways a name could be pronounced and then think of the spellings. Say the surnames out loud and spell them phonetically. Look for records under any of those spellings. Could Stone be spelled Stoan? Charette could mutate to Shorett, and Cayeaux could appear as Coyer. To help you think of potential spelling variants, examine the compilation of names from the 1790 census. As you say the names aloud, you may get new ideas for potential surname spellings you are researching.

TREE TIPS

Ask your friends and associates how they would spell the surname you are tracing, based on its pronunciation. Ask children, who may be more phonetic and unbiased in their approach to spellings than adults. You may come up with some new ideas on variant spellings.

Names were often corrupted as immigrants interacted with their neighbors. Perhaps the individual giving information to the census taker was not your ancestor, but a neighbor who happened to be home when your ancestor wasn't. The census taker took the information the neighbor gave rather than revisiting your ancestor, so Mr. Justice became Mr. Justis, Jean Christien became John Christian, and so on.

According to the 1850 census records, Diana Drollinger's grandfather was born in North Carolina about 1800. A search of the North Carolina census records and other documents turned up no Drollinger/Drolinger/Drullinger families. Although census information is sometimes incorrect, other sources also suggested a North Carolina birthplace. The key to the puzzle was in the microfilmed pension records at the National Archives. Henry Drollinger states in his pension deposition that his name is Drollinger, pronounced Trollinger in the German style. A check of North Carolina records for Trollingers uncovered numerous families, and eventually Diana's great-grandparents.

PEDIGREE PITFALLS

Never assume that the surname you are researching has stayed the same through the generations or even through a lifetime. How many times have you had to correct the spelling of your own name? Perhaps some of the misspellings will survive to someday confuse your descendants.

Different Record, Different Spelling

A name may undergo several transformations before it takes the familiar form you know. When the immigrant Weir/Weer accumulated enough money to buy a house, he recorded the deed in the county clerk's office where he became Wiere. Weir/ Weer/Wiere moved west and bought some land, and this time the clerk wrote Wear. Meanwhile, his account at the general store was under the name Ware, and that's the name the children learned to write when taught by the 18-year-old schoolmarm. So your grandfather is George Ware, but his grandfather was Samuel Weir.

Naming Patterns

Patronymics were widely used in forming surnames. This was used to sort out the individuals in a community, noting Richard's son, Thomas's son, and so on.

DEFINITION

Patronymics are names derived from a father's name or paternal side of the family. For example, if your father's name is "Tom," your surname may end up "Tom's Son" or "Tomson," which in turn might turn eventually into "Thompson."

Patronymics were particularly prevalent in the Scandinavian countries. For example, Mr. Jensen's sons all have the surname Ericssen. But as soon as you know that the surname is formed by adding *sen/son* to the first name, you can see how Eric Jensen's sons created the surname Ericssen. Daughters would have *datter* added to the father's first name, thus Eric Jensen's daughters would use Ericsdatter as their surname.

Other countries used the son suffix as well, such as Williamson or Jameson (also expressed as Williams or James, but did not change the surname with every generation.

Sometimes prefixes and suffixes were used to denote sons or daughters: *O*, *Ab*, or *Ap*; *Mac* or *Mc*; *Fitz*, *ich*, or *itch*; and *ev*, or *off*. The resulting names were sometimes altered, such as when O'Brien became Obrien or Bryan, or when Ab Owen became Bowen.

Several links to information on patronymics and other subjects pertaining to names can be found on Cyndi's List (www.cyndislist.com/names.htm), along with a list of helpful websites and links, including a series by Michael John Neil in *Ancestry Daily News*. (Another interesting link at Cyndi's List, shown as "Given Names," lists the 10 top male and 10 top female names from 1900 to 1977.)

Religious Naming Customs

In some religions, it is customary to name children for dead relatives in a specific order. In other religions, it is the custom to name the first boy for the maternal grandfather and the first girl for the paternal grandmother. In others, it is the custom to name the first boy after his paternal grandfather and the first girl after her maternal grandmother.

TREE TIPS

When working on a family you know to be a specific religion or nationality, learn the naming patterns common to that group.

Individual Naming Patterns

Some families devise a naming pattern of their own that persists for several generations. Perhaps the first male in every generation of the family is named Henry or James. Each generation perpetuates the tradition. When there are three or four first cousins all named William Smith, it may be difficult to sort them out, especially if they all remain in the same area throughout their lives. It may also be the family custom to name a child the same name as an older sibling who died in infancy.

PEDIGREE PITFALLS

Never assume that you have the correct person if there are two or more individuals in the same area with the same name. Gather the facts for each event to which they are connected and make comparisons.

Mother's Name Preserved

A mother's maiden name is frequently used as a middle name for either boys or girls. Other family surnames are sometimes used as well. Clark Stone Phillip's middle name is his paternal grandmother's maiden name. Mary Catherine Smith may be shown as Mary C. Smith in a record, but after she married John Jordan, she may be shown as Mary Smith Jordan or Mary S. Jordan.

If your ancestors were creative, you may have to be creative. Strange things happen in genealogy. For instance, a contemporary family has two girls and two boys. The boys carry their mother's maiden name as their surname because the mother is from a family with no boys and she wants the name to be carried on. The girls have their father's surname.

PEDIGREE PITFALLS

Don't assume that a child's surname-sounding middle name is a surname in your direct line. I once found a child whose name was Diana Courll Drollinger. Courll was not Diana's mother's maiden name, as originally thought. Instead, she was named for her married aunt: Diana Drollinger Courll.

Given Names Giving Us Trouble

It is not only surname variations that befuddle us, but also first names. Is Katharine Shimmin the same person as Catherine Shimmin? This name may also appear as Kathryn. First names may be Anglicized: Katharine may have been Katrina, Katja, Catherina, Katrintje, or Tryntje.

Five Children, Same First Name!

German children often had two given names, but went by the second or middle name. Frequently, all the boys in the family had the same first name or a variation of the same name, such as Johann/Hans, but were called by their second names. The girls in the family may have had the first name Anna or Maria. The children may have answered to both names or just the second. Different documents may use different names for the same child. These variations may cause you to wonder if you are dealing with one person or three, as you sort out Maria Elizabeth, Maria Christena, and Maria Caterina. This can be further confusing thanks to an adult returning to his or her first given name after being known by a second name as a child.

Today, parents-to-be thumb through books of names looking for ideas for the baby's name. Although there have always been trends in first names, your ancestors didn't rely on book lists. They thumbed through the Bible, or they named their children for friends and relatives or famous people. Hundreds of boys went through life carrying the first and middle name of George Washington or Benjamin Franklin.

The romance of the west inspired one set of parents in the naming of their children. Their little girl Sierra Nevada is enumerated as Nevada at one time, as Sierra another, and as Vada on her marriage license. As far as research has revealed, her parents never traveled west of Indiana.

He Was Called Billy; She Was Called Abby

Nicknames can throw you off the trail. If you always knew your great-aunt as Polly, you may be surprised to find that her name was really Mary. The Sally in a will may be the Sarah in a deed. William may be Bill, Billy, Will, or Willie. Bert may be a shortened form of Albert, Gilbert, Robert, and others.

Finding the name that the nickname stood for is not always obvious. This is especially true when the nickname is encompassed within a name, such as Gus for Augustus, Gum for Montgomery, or Fate for Lafayette. You may need to do some reading to know what names to look for. A particularly extensive resource is Christine Rose's *Nicknames: Past and Present.* The Connecticut State Library has culled some nicknames from their eighteenth- and nineteenth-century documents and put them online at www.cslib.org/nickname.htm.

Double Trouble

It has been said that everyone has a double, someone else in the world who looks just like him or her. The same thing is true of names. In San Jose, California, a man who had voted regularly for years was cut from the rolls. John P. Taylor called the Registrar of Voters and found that he had been cut because he was listed at two different addresses. The computer matched the John P. Taylors' months, days, and years of birth, found them to be the same, and deleted one from the rolls. But there were actually two different John P. Taylors with the exact same birth date; one was John Paul and one was John Phillip, but they had both registered as John P. Taylor.

Almost no one has a unique name. Names that are no longer common seem exotic to us, so we think the individuals will be simple to isolate. You may think that Tryphena Atherton should be easy to identify because, to you, the name is unusual. Early Vermont records have several Tryphena Athertons. You may think that the common name of Davis pared with Caleb is unique. In that case, you would be surprised to learn that there are at least four Caleb Davises in the 1840 census records in New York.

Your ancestors went by many names, just as we do today. Some have different names at different stages of their lives. Some women change their names when they marry; some do not. Siblings may spell their surnames differently. In some families, one woman may hyphenate her maiden and married names, whereas others may not.

Which Do You Pick?

When you begin to do genealogy research, you have the option of tracing any one of numerous surnames. How do you choose the one you want to work on? For many, this is a personal decision; for others, it's a matter of practicality. I began working on my maiden name because my father would never talk about his family, and I was very curious about who they were. As I began to gather information from many family sources, I became interested in other lines because I had more information and a better chance of success.

Some may choose a surname to follow because of a family tradition connected with it: "Our family is related to Kit Carson." "Our family is connected to President McKinley." Maybe you have a desire to prove you are a Mayflower descendant or the descendant of a Revolutionary War patriot. Maybe you seek admission to the Sons of the Republic of Texas. Perhaps you are eager to prove that you are a descendant of a Native American or a French trapper, or are eligible for a pioneer descendant certificate offered by many states.

Whatever surname you decide to pursue, one of the first things to do is to consider the different ways the name may have been spelled by your ancestors.

The Least You Need to Know

- Names are often spelled in a variety of ways.
- Ancestors may have translated or shortened their foreign name.
- Different branches of your family may use different variations of the same surname.
- Customs among ethnic and religious groups may affect the names of the children.

Getting the Most from Libraries

In This Chapter

- Finding the genealogical collections
- Using the catalogs and indexes
- Finding family histories
- Understanding different types of libraries

The previous chapter acquainted you with the diversity of written names. Now you're ready to tackle research in the array of books, websites, and special collections that can be accessed through libraries.

Libraries differ, as do their holdings. The small-town library has a limited budget to purchase books for special interests. However, if the local genealogy society is enthusiastic, their contributions, cake bakes, and book sales provide needed funds so that even small libraries usually have a few books on the subject. Check the surrounding area as well; you may discover collections in nearby towns or cities, or even a regional repository in an adjoining county.

The Bigger the Better

All states have a major repository. It may be a state library, a state genealogical society library, a state archives with a library, or a state historical society library. Some lucky states have all four. Each has its own strengths, perhaps extensive newspapers or manuscripts or an extraordinary number of family histories. Become familiar with all the collections in the geographic area of your search. Larger facilities should have an inventory or guidebook, or at least a leaflet describing important aspects of their holdings. Though the state repositories have larger collections, smaller local and regional libraries may hold magnificent treasures.

Make use of the Internet. Most libraries have websites, and some include digitized images, indexes, historical backgrounds, and other worthy enhancements to your family's history. For a directory of public libraries see www.publiclibraries.com or http://lists.webjunction.org/libweb/Public_main.html.

Making Your Way Around

You may be familiar with the standard card catalog that includes entries for books organized by subject, title, and author. Look also for catalogs of newspapers, schools, photographs, manuscripts, and other specialties. Most libraries now have *online catalogs* but many still offer their original card catalog onsite, even if they no longer update them. Sometimes the older catalog had important annotations not included on the newer online version.

While at the library, also inquire about special collections. Volunteers may have indexed all the obituaries before 1920 or transcribed tombstones from local cemeteries. Volunteers may have created a *consolidated index* of the names in the county histories.

DEFINITION

An **online catalog** is a database that users access by computer.

A **consolidated index** combines data from more than one source. For example, the indexes to five county histories may be consolidated into one.

Winding Your Way with Books

One fundamental book to seek out is Val Greenwood's *The Researcher's Guide to American Genealogy.* It devotes chapters to sundry categories of records.

Another important text is Loretto Dennis Szucs's and Sandra Hargreaves Luebking's *The Source: A Guidebook of American Genealogy* (the third edition), which covers record types in detail. Turn to this massive book again and again as questions arise and as you need more help. And don't overlook the third edition of the *Red Book* edited by Alice Eichholz, Ph.D. Each state has its own chapter that incorporates history and a useful discussion of types and locations of records, as well as courthouse addresses and data on the county's formation.

TREE TIPS

Seasoned genealogists refer often to in-depth guidebooks. These books are not meant to be read straight through, but rather to be read in doses as you encounter new situations or unfamiliar records.

Your Family on Its Pages

I was astounded the first time I saw a family history book among our family's books. I was so excited. Here was a book of genealogy devoted specifically to our family! My surprise turned into amazement when I visited my first genealogical library and saw that there were hundreds—no, thousands!—of similar books. These histories are usually devoted to one specific family or a combination of three or four families.

Because no collection is complete, check a number of repositories for a history on your family. Enlist your reference librarian's assistance. From your computer, go to the websites of major genealogical libraries and look for family history books by entering the surname into their website's search engine.

> **LINEAGE LESSONS**
>
> When you find a family history on the surname for which you are searching, whether in book form or on the Internet, don't assume that everything in it is correct. Being in print doesn't make it true. If there are no citations included so you can judge the reliability of the information, you'll have to do some independent research to establish that the history you've discovered is indeed accurate.

Catalogs to Help

Many major libraries (Newberry in Chicago, the New York Public Library, and others) have published catalogs. An example is the *Genealogies Catalogued by the Library of Congress Since 1986*, which contains over 1,300 pages and lists many hundreds of books. None of the catalogs contain every genealogy book published; they only list what's in their collection. Of course, they don't include any books published since the publication date of their catalogs, either.

WorldCat

Simply put, WorldCat (www.worldcat.org) is the world's largest network of library content and services. As touted on its website, its libraries are "dedicated to providing access to their resources on the web, where most people start their search for information." Their website lets you search thousands of local and worldwide libraries. If you are searching for a book, it helps you find one at a local library. Use it to find research articles, digital items, and other pertinent publications.

If you have an active membership with a library that owns the item you want to view, you can use a remote checkout and even have the book shipped or delivered to your home. Some advanced search features on their website are available only by accessing WorldCat through your library.

Magazines and Journals: Not at the Corner Newsstand

Family histories are not confined to books. Many periodicals and journals, particularly those national in scope, publish compiled genealogy articles. In some cases, these are extensive and may be serialized in several issues. Others may be limited, perhaps covering just the immigrant and one or two generations. Significant finds (proving the English ancestry of a Mayflower passenger, or the maiden name of the immigrant's wife) often appear on their pages. Libraries frequently have a special spot for newly arrived periodicals in the genealogy section. Spend some time examining your library's genealogy section to become familiar with what's available. Consider subscribing to those periodicals and journals of most interest to you. In some instances, you'll obtain a magazine subscription automatically when you join the associated organization.

LINEAGE LESSONS

In addition to the comprehensive periodical indexes, there are numerous consolidated indexes to specific periodicals, such as *The Virginia Genealogist* (no longer in circulation) and *New England Historical and Genealogical Register*. Many periodicals are digitized on CD-ROM with a searchable feature that takes the user right to the page, though some of the older CD-ROMs may not work with newer operating systems.

Crammed with Articles

How can you find published articles that may lead to information on your family? Consult the PERiodical Source Index (PERSI) and the Genealogical Periodical Annual Index (GPAI).

PERSI is published by the Allen County Public Library in Fort Wayne, Indiana. Their massive indexing project covers hundreds of periodicals in their library. Retrospective indexes in a number of volumes cover periodicals published from 1847 through 1985, with annual supplements since 1986. Articles are indexed by subjects, locations, and surnames. Look for PERSI in book form or on CD-ROM at your library, or access it online as an ancestry.com subscriber. You can also access it at libraries that subscribe to HeritageQuest. If you can't find the periodical with the article you want to read in your library, you can order a photocopy of the article for a small fee from the Allen County Public Library. Their library has a copy of every magazine they've indexed in PERSI because that index is built on Allen County

Public Library's own collection. A handy form for ordering copies of the article is available at acpl.lib.in.us/database/graphics/order_form.html. There you'll find a paragraph regarding the fee structure and a link to an order form you can fill out to request copies.

Though no longer published, GPAI (1962–2002) is still valuable for locating articles published prior to their cessation. GPAI indexed genealogies, lineages, Bible records, source records, and book reviews that appeared in various periodicals. Use both PERSI and GPAI to get broad coverage.

Not Always Bound

Libraries and State Historical and Genealogical Societies hold a variety of *original material*, some in the form of unbound manuscripts. In a visit to the Wisconsin Historical Society, I located a Bible record from an early New England family. The original pages, torn and deposited in the society's collection, were probably from a Bible that made its way west in one of the many migrations. It listed the father born in 1782 and the mother born in 1779. Included were their full birth dates, their marriage date, the wife's maiden name, their children, and much more. Letters written by early pioneers have made their way into similar collections. The business papers of a local merchant, a thesis for a college degree, original sheet music composed by an ancestor, or a genealogy chart sent from the old country—any of these and more are just waiting for your discovery.

DEFINITION

When referring to manuscripts, **original material** may be loose papers, letters, photographs, diaries, and other items. These materials may be maintained in a separate manuscript section of the library with its own catalog.

The Wonderful World of Reference

Don't stop at the library's genealogy room—keep walking, right into the reference section. Nationwide directories of funeral directors, newspapers (with addresses, date formed, and other important details), schools, and sundry others await your discovery. You'll need these as you write letters and check out what's online.

The reference section also holds multivolume biographical dictionaries such as the *Dictionary of American Biography* (New York: Charles Scribner's Sons, 1937, in 20 volumes, released as 10 volumes in 1946). Later, supplementary volumes were added covering the time period to 1980. Sometimes the reference has been digitized, either through a paid subscription website or a free university website.

These and other similar publications have marvelous biographies and pen sketches, and often signatures as well. Don't believe every word in them, but use them to guide you to supporting records.

In the following figure, this pen sketch and signature from Lamb's *Biographical Dictionary of the United States* (Boston: James H. Lamb Co., 1900) illustrate the marvelous reproductions you can sometimes obtain from biographical dictionaries. Their biographies often include details that help in tracking the families.

Pen sketches and signatures adorn the pages of the biographical dictionaries, gazetteers, and atlases.

Webster Was Never Like This

In the library's reference section are numerous dictionaries besides the standard *Webster's Dictionary*. Consider the multivolume *Oxford English Dictionary* when faced with archaic terms. This dictionary explains the many words and phrases no longer in use that you may encounter, including tools, implements, diseases, and others.

You will find an array of other specialized dictionaries on a wide range of topics. You can also go online and enter a term such as "archaic dictionary" in your Internet browser or favorite search engines for other choices. A few of the many specialized online dictionaries and lists are found at cyndislist.com under the category Dictionaries & Glossaries.

A Library Is a Library Is a Library

Wrong! The types of libraries vary considerably. Among those you may use are the following:

- Public libraries
- Private libraries
- Genealogical and historical society libraries
- Lending libraries
- University and college libraries
- Ethnic and religious libraries
- Lineage society libraries

Public libraries may be very small on the local level or very large on the state level. Most public libraries have materials you can check out, though that service may be restricted to residents only. Usually reference books, which may include genealogical material (or fragile or irreplaceable books), are not allowed out of the library. In private libraries, some collections are severely restricted in use. Usually, manuscripts housed in private libraries can be accessed by only certain individuals, though their librarians may assist you by mail. Many private libraries, however, do make their collections personally accessible to the general public.

Genealogical and Historical Society Libraries

The genealogical or historical society library may be public, or it may be the private collection of a particular group. If it is the latter, restrictions may apply and a small fee charged for its use. In some cases, members have *open stack* privileges, whereas nonmembers don't. If it's a closed stack collection, you must request the books and staff will retrieve the materials for you. There may be a delay (usually 15 minutes to an hour) before the staff member delivers the book to you. Most libraries that are considered closed stack have some of their more commonly requested books available on publically accessible shelves.

> **DEFINITION**
>
> **Open stack** refers to the use of the books. If the library has open stacks, patrons may freely examine books on the shelves.

Lending Libraries

Usually, genealogy libraries do not allow patrons to borrow the books, but there are exceptions. California's Sutro Library in San Francisco, for example, will allow some books out on interlibrary loan. Go to www.library.ca.gov and follow the links. The large collection at the St. Louis County Library in St. Louis, Missouri, (which now includes most of the volumes from the old National Genealogical Society Library of Arlington, Virginia) will lend many materials. Visit their website at www.slcl.org/branches/hq/sc/ngs/ngscol-main.htm. There are others. Ask the reference librarian in your own local library for assistance in ordering from other libraries. Borrowed materials are sometimes restricted to use only in the requesting library.

Though interlibrary loan is an advantage for those who can't travel, the disadvantage can be that the book you want to view may be out on loan when you arrive. To avoid disappointment, call the library before making a long trip. Sometimes loan privileges are based on membership to a society, while others may loan for a small fee. Some also lend microfilm or microfiche.

University and College Libraries

Among the least-used libraries in genealogy are those of the universities and colleges, though they hold some of our richest resources. In the Wise Library (previously the Colson Library) on the campus of West Virginia University in Morgantown, I located fragments of an original deed that had not been recorded. These fragments indisputably proved the parentage of a Virginian born in the mid-1750s. See www.libraries.wvu.edu/wvcollection for information on this invaluable collection. At the Bancroft Library at the University of California in Berkeley, I was thrilled to hold an account book written in the 1700s giving information on the Thomas Jefferson family heretofore unknown by many researchers.

TREE TIPS

When visiting a university library, inquire about their manuscript catalog. That's the most likely place to find the "hidden" treasures. Search it by surnames, localities, and subject.

Almost all universities have websites describing their collections. You'll discover extraordinary websites such as "Making of America" (MOA, http://quod.lib.umich.edu/m/moagrp), a joint effort of the University of Michigan and Cornell University. MOA brings us a digital library of thousands of volumes. It offers an incredible array at no cost.

Want to learn more about surveying by our government? Check out J. H. Hawes' *Manual of United States Surveying*, published in 1873. Been fascinated by items you've found referring to the early minister Cotton Mather? Check out his *Magnalia Christi Americana* at the same site. Social histories, atlases, and much more are available there.

For a number of other books at the MOA site go to http://quod.lib.umich.edu/m/moa.

Locate the *Official Records of the Union and Confederate Navies in the War of the Rebellion (1894–1922)* at http://digital.library.cornell.edu/m/moawar/ofre.html. For *The War of the Rebellion: A Compilation of the Official Records of the Union and Confederate Armies (1880–1901)* go to http://digital.library.cornell.edu/m/moawar/waro.html. This is just the tip of the iceberg of exciting finds at university websites. Collections such as these, available with a few strokes of our fingers, awaken us to the power of the web to truly put flesh on the bones of our forebears. To put even more flesh on bone, search Google Books for information on social history, "current events," jobs of the time, topography, and geography.

FamilySearch/The Family History Library

The Family History Library (FHL) is maintained by the Church of Jesus Christ of Latter-Day Saints in Salt Lake City, Utah. As a result of their religious convictions, their genealogical collection is immense and worldwide. It isn't limited to use by members of the church, either; the materials are available for anyone to use. Their library catalog and other computer projects make access to many of their records relatively easy. They maintain hundreds of FamilySearch Centers across the country. Anyone can order microfilm for use at these branches by going to their local FamilySearch Center.

FamilySearch.org is an enormous collection of several genealogical databases, lessons consisting of videos and written papers, and much more. One of the most useful segments of FamilySearch is the Family History Library Catalog, with its listings of their extensive microfilm holdings. Most microfilm rolls can be borrowed for a nominal fee through your local FamilySearch Center. You will find microfilm for deeds, probate records, vital records, civil records, and much more, worldwide.

TREE TIPS

To locate a FamilySearch Center near you, go to www.familysearch.org/eng/library/FHC/frameset_fhc.asp or check your phone book.

See Chapter 5 for more about familysearch.org.

Other Magnificent Collections

The Allen County Public Library in Ft. Wayne, Indiana, houses one of the largest collections of genealogy material in the United States. For years, they have systematically added to their holdings, which are considered some of the finest in the country. Their strong collection of periodicals resulted in the development of the PERSI index, described earlier in this chapter. Additionally, thousands of genealogy books are available on their shelves. Access their catalog at http://smartcat.acpl.lib.in.us/?skin=genealogy&q=Search+the+ACPL+Catalog.

Other major collections include the New England Historic Genealogical Society Library in Boston, Massachusetts (www.americanancestors.org/about), the New York Public Library in New York City (www.nypl.org), the Wisconsin Historical Society Library in Madison, Wisconsin (www.wisconsinhistory.org), and the St. Louis County Public Library in St. Louis, Missouri (www.slcl.org). Their already extensive collections continue to grow as interest in genealogy escalates.

Religious and Ethnic Libraries

Some churches and ethnic groups maintain libraries and archives. Their hours may be limited, so inquire ahead of time if you plan to visit. Check with your local reference librarian for guides that will lead you to the location of many church archives. Also go to www.cyndislist.com and select the category Religions & Churches. Alternately, insert the name of the denomination in your browser, and add the word "archives" for listings. Enormous amounts of information exist on websites—general background on the religion, descriptions of their library holdings, and perhaps even actual images of records. Here are a few of the websites:

- Baptist: The Southern Baptist Historical Library and Archives at www.sbhla. org. Included is a list of microfilmed church records available for purchase, and they have a photo gallery and brief biographies.

- Catholic: Links to websites of dioceses and archdioceses worldwide are available at www.cyndislist.com/catholic.htm.

- Huguenot: The National Huguenot Society's website can be accessed at www.huguenot.netnation.com/general. The Huguenot Society of Great Britain and Ireland is at www.huguenotsociety.org.uk/resources. These sites include such informative material as "Who Were the Huguenots" and "Important Dates in Huguenot History," addresses of collections, and other aids.

- Jewish: The American Jewish Archives at www.americanjewisharchives.org/ intro.html and "Reading Hebrew Tombstones" at www.jewishgen.org/ infofiles/tombstones.html are both helpful.

- Lutheran: The Evangelical Lutheran Church in America has a website at www.elca.org/library/library.html. The Lutheran Archives Center at Philadelphia can be accessed at http://ltsp.edu. These sites include links to other sites.

- Mennonite: The Mennonite Archives is available at www.mcusa-archives.org. Included is an amazing collection of "Mennobits," obituaries of thousands with a search feature. Try also www.ristenbatt.com/genealogy/mennonit.htm for background and links, and even tombstone inscriptions from a number of cemeteries.

- Methodist: The United Methodist Archives at www.gcah.org is one of the sundry Methodist research sites available online. The Methodist Archives and Research Center (England) at the John Rylands University Library of Manchester is in the special collections at http://library.cmsstage.manchester. ac.uk/specialcollections/collections/uomarchives.

- Presbyterian: Presbyterian Historical Society at www.history.pcusa.org is a good place to start; once there, go to Collections. There's also some good information at their Presbyterian History link.

- Quaker: The Friends Historical Library of Swarthmore College in Pennsylvania at www.swarthmore.edu/fhl.xml includes explanations of Monthly Meetings, description of the kinds of records Quakers keep, which monthly meeting to search, abbreviations, The Quaker calendar, and more. Many other Quaker websites are also available—enter "Quaker" in your browser and off you go!

Links can also often be found to lesser-known religions—use a search engine if you find mention of one in connection with your family and you may be able to learn more.

Lineage Society Libraries

The best-known and most extensive library in this category is the National Society Daughters of the American Revolution (NSDAR) Library in Washington, D.C. Members of the NSDAR can use the library at no charge; others pay a small daily fee. It has considerable material, including Bible records submitted in support of applications. Go to their website at www.dar.org. An awesome database is available on their website at http://members.dar.org/dar/darnet.cfm; it indexes thousands of records in their Genealogical Records Collection (GRC). Use the search engine provided after you've read the "Index Overview" on the website. Check also Eric Grundset's *American Genealogical Research at the DAR*—it provides many tips for successfully using this library.

TREE TIPS

Most lineage societies have a members only policy. If you are interested in joining and you qualify, an existing member can sponsor your membership. Most lineage societies require extensive proof of relationship accompanied by source citations, so prepare carefully.

Besides the well-known DAR Library, there are also others. Read the pertinent chapter in Ancestry's *The Source* or check www.cyndislist.com for lineage societies.

The Library of Congress

The Library of Congress has a genealogy department, but don't stop there. Look for their rare books section, the newspaper collection and finding aids, the photographs collection, and the map section; all will likely be important to your search. Consult James C. Neagles' *The Library of Congress: A Guide to Genealogical and Historical Research* for a better understanding of the vast facilities.

Don't overlook the Library of Congress's extensive website at loc.gov. There is much of interest here, but be sure to click on American Memory for a seemingly endless well of genealogical treasures: photographs, images, sounds, background information, taped interviews (such as those with many former slaves), maps, and much more.

The Least You Need to Know

- Genealogical collections and repositories vary drastically in size. Most have websites.
- Many family histories are published in book form or posted on the web. Verify them if proper citations aren't included.
- Libraries have more than books; manuscripts, maps, newspapers, photos, and multimedia can provide crucial information for your search.
- University and religious libraries are an excellent and often overlooked resource.

Your Family's Hometown

In This Chapter

- Using the public library to find relatives
- Learning to recognize the records your family left
- Using county records
- Finding family treasures in museums

You may be one of the fortunate people who live in the same county where your family has resided for more than one or two generations. Within your family's own county, there are a number of repositories to assist in your search. You can become familiar with the records your family may have generated in the county in which they lived. If your family didn't live in the same area in which you now reside, consider making a trip to where your parents and grandparents lived most of their lives. This chapter concentrates on what's available in the libraries and museums of your ancestors' hometowns.

Chapter 14 tells you how to prepare for a trip to your ancestors' hometowns, and Chapter 15 explains the materials available in the courthouses there.

Go to the public library where your immediate family lived. Ask if there's a genealogy section and a local history section; both will be useful. The genealogy section may include family histories, books devoted to the genealogy of a specific family, and other published and unpublished manuscript aids related to lineages. The local history section consists of books, pamphlets, scrapbooks, and other items pertaining to the county. Find out if there's a genealogical society in the area. They may meet in the public library, but they often have their own headquarters and library.

Some of your family's records will be in their original form in the courthouse. Some may have been abstracted and published—those you will find in the library. The library may also have original documents—scrapbooks, photographs, and more. In some cases, even "discarded" courthouse records have made their way to the local library collection

TREE TIPS

Ask at the library for the name of the official county or city historian. These historians can point you to local collections, and they may even be the custodian of some of those records.

Getting Started in the Public Library

Take with you the names, dates, and locations of your family members who resided in the county. Your goal is to find their birth dates, marriage dates, spouses, and death dates (for starters). Be on the lookout for records that mention the churches they attended or their religion. (Knowing that, you can later check to see if the church has any records naming them.) A record showing where they were buried can supply you with leads. Each major event in their lives that has associated records may help identify their relatives and reveal snippets of their lives.

Your purpose is to find everything that may bear on your family. Nothing is too insignificant to note. Write down fully in your notes the day of your library trip, the name and address of the one you visited, and in which book or file you found the information so you can find it again. In library holding, you may learn the following:

- Birth date and birth place

- Names of parents

- Death dates and town or township

- Marriage date, place, and spouse

- Biography, possibly with a photograph or pen sketch

- Military service

- Names and addresses of relatives and when they moved into or left the county

- Occupations of parents and relatives

- Acreage owned by family members

Initially, you may be unable to locate something specific. Stick with it. As you learn more about the records, your successes will increase. With each success, your enthusiasm for the search will increase, too!

TREE TIPS

To access an alphabetical listing of public libraries nationwide, go to publiclibraries.com

Shelves of Possibilities

After you start browsing the library shelves, you'll begin to discover the many available resources. Books and finding aids of every description exist. Look for such items as the following:

- Old Settlers' files
- Obituaries/necrology/funeral records
- Tombstone surveys
- Vertical family files
- County histories and indexes
- Published vital records
- Published deeds, probate and court records
- Scrapbooks
- Gazetteers and atlases
- Voting registries
- City directories
- School records
- Artifacts and photographs

Old Settlers Remembered

Many counties have projects geared toward preserving information on their earliest settlers in what are commonly called "Old Settlers'" files. These may consist of recollections of descendants, biographies, and even taped interviews. Examine them for any of the surnames of your family who resided in the county. Note also the name and address of anyone who submitted information; they may be alive or have family members still living. Contact with them may yield some wonderful memorabilia. Just remember not to accept what's written as fact until you've verified it.

Death Records

Many local libraries have indexed their early obituaries or established a *necrology* file. The content of the obituaries varies. Normally, those in smaller towns were more extensive than in larger cities. Check this file for members of your family. (See Chapter 17 for more on necrology files and obituaries.)

DEFINITION

Necrology is a list of people who died within a certain time frame, or a collection of obituaries. You may find the collection so titled in the library.

Vertical Files: Just Waiting to "Talk"!

As interest in genealogy intensifies, libraries receive letters of inquiry from all over the country. To preserve these, they often create "vertical files," a set of folders stored in a filing cabinet, usually by the family surname. Included may be letters, clippings, Bible records, photographs, research notes, charts, and others. These can yield new clues, and equally as important, may supply the names of others seeking your family's history.

County Histories: The Mug Books

Almost every county in the United States has had at least one county history published, and some have several; they vary in content. Those published in the late 1800s or early 1900s typically consisted of brief histories of each of their townships or towns, churches, lodges, medical profession, schools, newspapers, county government, and notorious happenings, and even those who served in the military from the county. They may include the name and place of origin of the early pioneers of the area—who established the first grist mill, the first physician, town officers, and other similarly valuable information. Consult P. William Filby's *Bibliography of American County Histories*. In your Internet browser, insert *bibliography county histories*, and listings for some specific states will likely be displayed. Look in your library for town histories. Though town histories aren't listed in Filby's bibliography, many have been published.

The county histories are sometimes referred to as "mug books" because often they included biographies with photographs (or pen sketches) of the early citizens. The lack of a biography was no reflection on a person's standing in the community, though; the books were mostly on a subscription basis. Those who paid were included; those who didn't pay weren't in them. To subscribe or not depended upon the frugality and monetary priorities of the individual. The histories published in the late 1800s were supplied by people far closer to the time of the events and should be (but aren't always) more accurate than the recollections of present-day descendants.

There are some websites that may lead you to the county histories you seek. Try genealogybranches.com/countyhistories.html. Many county histories can be found at Google Books (books.google.com). It's also possible to find a county history that's been digitized and is now offered on CD through either small publisher websites or on eBay (eBay.com).

PEDIGREE PITFALLS

Note the year of the book's publication; was your ancestor living then? If not, someone else provided the data and may have guessed at some of the facts.

View the mug books with caution. If the biography includes several earlier generations, there may be multiple errors caused by memory loss or a lack of knowledge of the family background (or even, sometimes, by a desire to elevate their standing in the community or to obscure details of a less-than-desirable past). You must confirm all details. Nonetheless, the biographies are unique and provide an insight often lacking in any other source. You'll learn of your ancestors' schooling, jobs, purchase of the farm, the churches they attended, when they "found" religions, and other fascinating facts.

TREE TIPS

When you are faced with finding the family in an unindexed book, try to find from other sources the community or township in which they lived. Then you can start your search in that unindexed book with its table of contents mentioning localities.

A common problem with published county histories is the absence of an index. Usually, only the name of the subject in the biographies was listed; other names within the sketch were not. Rarely were the names in the town and historical sections indexed. It can be a tedious process to locate your ancestor's name among the pages. Fortunately, many individuals or groups initiated projects to remedy this shortcoming. Even if an index hasn't been published, there may be a card index or a computerized print-out at the library created by locals for the histories of their own county.

Vital Records: The Facts of Life

Vital Records—births, marriages, and deaths—are among the richest of documents that help build the family tree. Searching through them may take considerable effort because it often requires going to a variety of locations. They can be scattered among the shelves of the courthouse, city hall, county health department, local historical society, and even church and state archives. First, check the library to see if anyone has compiled or published any of the local vital records. Although you don't want to rely upon the published version (because of possible omissions and errors in transcriptions), you can use it to initiate the search. The preface of the book of vital records may explain where the various records are housed, available time periods, and which vital records, if any, have been destroyed.

The Web Shines Here

The Internet provides a variety of ways in which you can supplement the search for a county's vital records. One valuable website is vitalrecordsus.com. You can also go to cyndislist.com, and select the category "Death Records" (or any of the other listed vital records).

Another alternative is to go to usgenweb.org and use their links to get to the state and county of your search. Examine the county website carefully to see if any of the vital records have been posted, published, or microfilmed.

A nifty website is deathindexes.com, which provides searchable death indexes and records by state. It also includes obituaries, probate indexes, cemetery and burial records, and information on the Social Security Death Index (SSDI). Some states have started putting historic death records (including PDF files that show the original document) online.

Local Records

If you're fortunate, your county may be one for which abstracts of other early records have been published. These might include an assortment of land records: deed abstracts, surveys, land entries, and others. Though you should also examine the originals in the courthouse, the published records have the advantage of an all-name index for the book, whereas the courthouse clerks index only the main parties.

A multitude of other published abstracts may exist: court minutes, order books, wills, inventories, and more. Look for mention of members of your family and clues to relationships. Later, make your first trip to the courthouse and experience the excitement of using the original records. For now, the published books may aid you in understanding the variety of available records.

Local Scrapbooks

Preparing scrapbooks of local events was a pastime for some townspeople; it gave them a sense of community. Reading these scrapbooks will give you a sense of life in the community as it grew. Though seldom indexed, the scrapbook pages can reward you with one or more articles involving your ancestor.

TREE TIPS

Can't find a location in the United States? Go to http://libguides.asu.edu/content.php?pid=10928&sid=79050 for several links to sites dedicated to helping researchers locate place names.

Gazetteers and Atlases

The county gazetteer (a geographic dictionary) or atlas, published in the late nineteenth or early twentieth century, is a treasure. The pen sketches of the homes and farms are priceless. The wagons and farm implements in the yard, the crops growing—these reflect a way of life. Also included are township maps, many showing the name of the property owners on their section of land. Nearby cemeteries may be noted as well. These not only help in establishing the family's residence, but they also may lead to the location of family burials. Acreage, occupation, and even their place of nativity may be included. The names of the neighbors may also be helpful.

The Historical Atlas Map of Santa Clara County, California, published by Thompson & West of San Francisco in 1876, includes a Business Directory. Businessmen are listed by name, address, occupation, nativity, year they entered the state, when they came to the county, their post office, and the number of acres owned. In it, you can learn that Sheriff J. H. Adams came in 1849 from Illinois during the gold rush, and that Dr. Benjamin Cory of Ohio came even earlier, in 1847. Don't overlook these wonderful atlases. Some pen sketches measure 12" × 14" whereas others are 8" × 12" or other sizes. They are unique in design and historical significance. Some of the commercial companies, such as Thompson & West, Lewis Publishing, and others, specialized in such histories, preparing them for many counties and states. The originals are collector's items.

If you can find a pen sketch of your family's home or farm in an early atlas, a good reproduction of it would be wonderful for framing. The following figure shows the office and residence of G. W. Breyfogle, M.D., at the corner of Third and St. James Streets in San Jose, California. This reproduction from *The Historical Atlas Map of Santa Clara County, California*, provides a wonderful sense of a bygone era.

The office and residence of one of San Jose, California's early physicians.

On cyndislist.com select the category "Maps, Gazeteers & Geographic Information" for numerous items of interest. Also check out the Library of Congress "American Memory" page at http://memory.loc.gov/ammem/index.html.

He Voted with Pride

Some areas may have published their early voting registers. In California, the published registers of many counties before 1900 are in libraries. They include data not easily obtainable from other sources. The *Great Register for Santa Clara County* for 1892 includes age; physical description (height, complexion, color of eyes, color of hair); visible marks or scars, if any, and their location; occupation; country of nativity (usually shown as the state); place of residence; post office address; date when naturalized; place where naturalized and by which court; date of registration; and whether sworn.

Sometimes volunteers have transcribed the whole of an early voting register. Check out the great site for Yuba County, California's voting register at yubaroots.com/registers/registers-index.htm; Included are the years 1872, 1876, 1877, 1879, 1880, 1886, 1888, 1892, 1894, and others, all easily accessed using the site's search feature. You'll find entries for those whose nativities included England, Ireland, Scotland, Germany, and virtually every state.

A marvelous sense of the man—what he looked like, how he made his living, if he was an immigrant—are all revealed in these registers. John Cotler, a farmer, was age 58 and born in Ireland. He was naturalized 17 August 1855, in the U.S. District Court in Boston. His descendants now know where to find his naturalization papers. Charles Cranz, age 74, was born in Germany, naturalized on 10 April 1840, in Canton, Ohio, and then lived in California. You learn, too, that he was 5' 7" tall, with a light complexion, blue eyes, and brown hair, and that he was a farmer by occupation. Though not all states have such detailed registers, determine what's available for the county of your search.

Voter registration information varies from state to state. More current registration rolls have residence and telephone number, and may have party affiliation. Accessibility to voter registers depends upon the laws of the state in which you're researching. A few voter registrations are listed at www.cyndislist.com\voters. You may achieve more success by typing *voter registration archives* in your browser to bring up others such as the 1867 Alabama voter registration images at archives.alabama.gov/voterreg/search.cfm, and the 1888 Chicago voter registrations with images at ancestry.com. Maybe you'll get lucky and find a listing like this one: Thos. Rowe, 477 5th Ave., born Ireland, white, in the county and state 18 years, and listed as naturalized.

In 2001, ancestry.com added the California voter registers for 1866-1898; a wonderful resource.

TREE TIPS

If you're unable to learn about voting records at the library, check with the county's registrar of voters.

City Directories: Home Sweet Home

The availability of city directories varies. Larger cities, such as Philadelphia, Boston, Baltimore, and New York, have directories extant from the early or mid-1800s and sometimes even earlier. Small towns may at first be included in the directories of their neighboring cities but they probably had their own in later years. Very rural and sparsely settled areas may have all the towns in the county in one volume.

Want to know if there is a city directory online? Go to citydirectoryrecords.com. Remember, no directory listing online is complete—find two or three others, too; this one will get you started. Some subscription sites also have selected city directories.

It's unlikely that within a county, the city directories are all housed in the same repository. Mid- and large-size communities each have their own library and each maintains a set of its own directories. After you establish where in the county your family lived, determine if the town has its own collection. These directories will not only place your ancestors in the county or town at specific times, but they also may lead you to the old home, perhaps still standing.

Additionally, watch for the following:

- Others of the surname listed at the same address; they're relatives.

- A widow's listing, giving her deceased husband's name and occupation.

- An individual's first listing in the town, to indicate arrival or coming of age.

- An individual's last listing in the town, to indicate a departure from the area or death.

If the directory has a reverse listing by street address (called a householder's index), check for neighbors. Some may be married daughters (families often lived in close proximity). After you learn their occupations, you can follow your ancestors from job to job. You may even be able to pinpoint the date of death within a year or two if the husband and wife are listed in one directory, and then in subsequent directories only one appears as a widow or widower. If the whole family disappears from the directory entirely and doesn't reappear in subsequent issues, they probably moved from the area.

Another useful section in the city directories is that of businesses. If your ancestor was a tailor, examine the business listings of tailor shops. Also examine the advertisements. They're charming, and you may be rewarded with your ancestor's ad entreating the public to purchase the finery he or she offers.

LINEAGE LESSONS

Don't assume the spouse died in the same year the surviving spouse is first listed as a widow or widower. It may take a year or two for the directory listings to reflect such events. If a husband is shown in several editions with his wife, and then is shown alone, he is likely widowed (though they could be separated). Watch for subsequent issues that may reveal the name of a new wife.

Learning with School Records

If you know what school your family member attended, and it's still in existence, contact them for school records. They may be reluctant to furnish records from their files, but sometimes they will if sufficient time has elapsed. If the school is no longer standing, try the county's department of education. If you don't know which school they attended, examine the city directory for the appropriate time period and determine which schools are listed on nearby streets.

TREE TIPS

To locate schools that may have existed in your family's neighborhood, obtain a detailed street map of the town and use it together with the city directory and telephone book.

She Grew Roses; He Went to Lodge

Clubs, lodges, and fraternal organizations may still hold records of your family. Some have national headquarters and will answer mail inquiries if the sender includes a self-addressed stamped envelope (SASE). The published county history, discussed previously, may give some information on the organizations that existed when your ancestors first lived in the area. Determine if any are still in existence. Some may have websites and email addresses. If your ancestor was a member of the Chamber of Commerce or another civic organization, get in touch with those organizations. Even if they don't maintain records of past members, they may remember your family and offer some recollections. They may have retained newsletters or minutes of their organization that can add pizzazz to the life history you're building.

The women may have joined garden clubs, knitting circles, reading circles, and church-affiliated groups. Try to find those groups; doing so will give you a glimpse of your relatives' personalities to know their interests and hobbies.

Artifacts and Memorabilia

Museums can vividly depict how your ancestors lived. Relics from the early times of the community—photographs, sketches, portraits, and more—assist in portraying your ancestors in the community. If there was musical talent in your family, you will enjoy seeing the old instruments they played. If your great-grandfather was a druggist, some of his own paraphernalia may be included in the museum's display.

Be sure to inquire about indexes to manuscripts or photographs or other holdings of the museum. If your ancestor was a collector of anything—postcards, thimbles, or fans—see if the collection was donated to the museum. Such collections are often annotated with intriguing bits of information about the donor. When the Arizona State Historical Society wrote to me to inquire about a quilt in their collection donated by a descendant of the town's jailer (related to my husband), they were able to supply a fascinating tradition of the quilt being stitched by the girls at the Bird Cage Saloon in Tombstone, Arizona. It was said to be their gift to the jailer's baby daughter.

 TREE TIPS

For a listing of hundreds of online museums in the United States, go to http://icom.museum/vlmp/usa.html.

Photographs: A Peek into the Past

Perhaps the most treasured finds in local museums are the photographs. Families who long despaired of ever finding a likeness of their great-great-grandfather may find his class photograph framed and displayed in a case. Museums normally have indexes to such collections and can assist you in locating them. Consider paying them to have a copy made.

The Library of Congress' "American Memory" collection at http://memory.loc.gov/ ammem/index.html has some extraordinary photographs, as do many state library and museum online databases.

The Least You Need to Know

- The library in your ancestors' hometown has numerous manuscripts and published records.
- If your family lived in your own town, you'll find vital records, court records, funeral records, and much more in your own backyard.
- Additional records for search include school records, city directories, and voting registers.

Following the Trail

It's time to start searching for your family in the census. Through it, follow them in their journeys as they migrated north, south, east, and west. The thrill of finally finding your family in the census records, and realizing probably at least one of them was there at the very time the census taker was writing down the information, will make you feel very close.

You may be like many others: You hate to write letters or emails. How to start? What to ask? This'll be easier with the guidance you read in this part. You'll find tips to increase the power of your letters and Internet inquiries, making the most of every one of them.

Have You Done Your Homework?

In This Chapter

- Getting to know the census
- Preparing for records research
- Understanding the Soundex code
- Coding your surnames

Most genealogists start their records research with the U.S. federal population census, a count taken every 10 years since 1790 to determine the number of congressional representatives for each state. Our forefathers didn't know they were laying the foundation for one of the most fundamental research documents for genealogy.

Census Importance

Why are the censuses so important to your quest for information on your ancestors, and why are they among the first records to search? Widely available, they place your ancestors in a specific place at a specific time, and the related information on them leads you to other locations and records.

The purpose of conducting a census is to account for every individual living in the United States on a designated day. The chances are good that your ancestors were enumerated if they resided in the United States on the day of the census. In later chapters, I'll make suggestions to help you find them if they don't seem to appear in the census data.

LINEAGE LESSONS

Even if an enumerator doesn't visit every residence in his or her area on the date deemed census day, the enumerator is to collect the information as if he had done so on the designated day. Let's say the census date is 1 June, and the enumerator doesn't get to a household until 13 June. The enumerator wouldn't list a baby born 20 June in the census record because the baby wasn't in the household on the census date. Similarly, if an individual died 2 June, the person would be listed in the census record because he or she was alive on 1 June, the designated date.

What can you learn from the census? Over time, census questions have pertained to military service, citizenship, marital status, and other topics. Some of your family members' responses may surprise you. Thinking the family 100 percent southern, you may be flabbergasted to find out your great-grandfather was a Union veteran from the Civil War. If family tradition says your third great-grandfather was born in Ireland, but the census information says he and his father were born in Virginia, then you have a discrepancy to check.

The census records may be the first place you find your ancestors in their family groupings. Maybe your grandmother is older than you thought, or perhaps your mother has an older brother no one mentions.

Preparation Saves Frustration

Before you log on to your computer or head out to a repository to search the census records for your family, do your homework. Make a list of the likely heads of households (the person in charge of the family unit, such as a husband/father, grown son, or widow) for whom you have gathered some information either by talking to your relatives or going through all the material you found at home, or both. Be sure to include any variant spellings of the surnames.

For each individual on your heads of households list, include a time period (the estimated dates of his or her lifespan) and a probable state and county of residence. For example, take a look at the following sample of a census search:

Abraham Gant	1930	Morgan Co., KY
	1920	Morgan Co., KY
	1910	Estill Co., KY
	1900	Estill Co., KY
	1880	Orange Co., VA
Joseph Jaspers	1930	Monroe Co., NY
	1920	Monroe Co., NY
	1910	Monroe or Erie Co., NY
	1900	Erie Co., NY
	1880	too young

The approximate birth dates of the individuals on your list will provide a starting point for the census search. (Note: I've omitted the 1890 census from the list because it's virtually nonexistent.) Which individuals would likely be in the 1930 census records? Which ones would be in the 1910 census records? If your parents were children in 1920 or 1930, they would be enumerated with the people they were living with at that time: parents, grandparents, aunts and uncles, or other.

Unless you suspect otherwise, assume initially that any ancestors under age 21 were living with their parents, and start your research with those parents. Group your list of individuals by the areas where you expect they were living.

Where Did They Live?

The county is a division of government. It's wise to begin looking for your ancestors in official records, so first determine in which counties your ancestors' towns belong. If their town has disappeared from current maps, look up the town in a gazetteer. There may be several towns of the same name; be sure to get the one in the area where your ancestors lived. You can usually find gazetteers in your state or university libraries. An 1895 United States gazetteer is available online at livgenmi.com/1895. For more resources, check Cyndi's List (cyndislist.com) under "Maps, Gazetteers & Geographic Information."

TREE TIPS

Sanborn Fire Insurance maps can add considerable interest to your family's story. Go to lib.berkeley.edu/EART/snb-intr.html to read about these.

The more accurately you can locate your ancestors' residences—whether street address, township, or ward—the more quickly your search will likely go. Don't be discouraged if those are the very things you're hoping to uncover by searching the census. You can be successful in your search; it just may take a little longer. If you're unable to zero in on a county, the search is not hopeless. Examine carefully the state census indexes for your ancestors' surnames. Often, you can determine the county from those indexes by identifying where there are clusters of the surname. The addition of searchable online census indexes makes it possible to sometimes find an ancestor even if you don't know the state. If the surname is common, however, you may get too many returns to narrow the information down to a reasonable number to check.

TREE TIPS

Review the material you've already collected about your ancestors. You may find the county of residence among the family papers, on death certificates, in obituaries, or in city directories.

Which Census to Search First?

Because of privacy laws, the latest census available for research is the 1930 census. (The 1940 census is scheduled for release on 2 April 2012, so you may have access to the census records by the time you read this.) Generally, you should start your research with the most recent censuses—1930, 1920, 1910, or 1900—and track the individuals back through the censuses taken during their lifetimes. The objective is to conduct a complete census search for each individual on your list. You may wonder why you need to keep getting additional census records when you found them in one census listing. There are three main reasons: (1) you need to compare the data you find to the data you collected previously, (2) the composition of the households may change from census to census, and (3) each census record has different information that may lead you to other records.

Last Touches for Your List

Your initial research list should now be a list of individuals with some indication of the census years in which you may find them listed, the probable counties and states in which they lived during those years, and the most recent available census records in which you might expect to find them. At this point, to prepare for your search for the actual census enumerations, it's time to code the surname for each individual you'll be searching for in the censuses of 1930, 1920, 1910, 1900, and 1880. (The 1890 population census records are virtually nonexistent. Nearly all were destroyed by a fire on January 21, 1921, in the Commerce Department building.)

Cracking the Code

Federal indexing for the 1880, 1900, 1910, 1920, and 1930 censuses is based on a phonetic system called the *Soundex* (or a similar one called *Miracode*). It was devised to overcome the vagaries of spelling by grouping together surnames that sound alike but have multiple spellings. In the Soundex, Bream, Breem, or Briem can be found in indexes under B650; Wier, Weer, Wiere, and Ware is coded as W600.

> **DEFINITION**
>
> **Soundex** is an indexing system based on the phonetic sound of the consonants in the surname. Each name is assigned a letter and three numbers. The letter is always the first letter of the surname. The **Miracode** for the 1910 census uses the same sound system, but the resulting lists appear in order of the visitation number assigned by the enumerator, rather than in order of page numbers of the census schedule as in the Soundex.

The 1880 census was Soundexed only for households containing children under the age of 10. (These children were the first to become eligible for old-age benefits in 1935.) This doesn't mean other households weren't enumerated. Rather, it means that if your ancestor wasn't sharing a household with children under the age of 10 on the census date, he or she wouldn't be in the Soundex indexes; he or she would, however, appear in the census record itself.

Those 1880 Children Retiring

After the Social Security Act of 1935 passed, the government had to determine who was eligible for benefits. Those eligible needed to prove their ages. If they had no birth record, they could help substantiate the birth by a census record. The government, realizing applicants in the first group were born before statewide birth registration, needed an efficient way to locate individuals in the census records for verification of birth. The only ones in the household that were important to the government in this initial Soundex were those who were under age 10 in 1880. It was that group who would later be applying for Social Security.

Using the Soundex System

Before census records were readily available online, it was necessary to "soundex" a name in order to use the 1880 federal census soundex on microfilm. Most researchers now use the readily available online indexes and digitized images and no longer need to rely on soundexing the surnames to conduct a search. If you plan to do a census search in microfilm instead of searching online, then you need to understand how to soundex a surname. Read the instructions for how to Soundex names at archives.gov/research/census/soundex.html.

Shortcut Coding

Many genealogy programs will generate the Soundex codes for every surname in your database. Also, there are Soundex code generators online. You can enter a surname and the generator produces the code for you. One (http://resources.rootsweb.com/cgi-bin/soundexconverter) is a RootsWeb page that will return not only the code for the surname you enter, but also a list of other names with the same code. Some online code generators such as stevemorse.org/census/soundex.html return the surnames with both the Soundex and Daitch Mokotoff Soundex codes.

Soundex Codes and Online Indexes

Although Soundex codes aren't necessary to use online census indexes, understanding the Soundex and applying it to the surnames on your list may still prove important. You may be unable to find your surname with a simple alphabetical search, in which case you'll need to use the Soundex code search option. Other online indexes, particularly those for ships' passenger lists, use Soundex codes to increase the chances of success in finding the person whose name wasn't spelled the way you thought it was. If that happens, go to the suggested NARA website mentioned previously and study the system.

Indexes Before the Soundex

There are various published indexes for the 1790–1870 censuses before the first Soundex of 1880. Although most indexes cover an entire state, some are regional. Check for indexes at all the libraries and repositories where you look for sources.

Consolidated comprehensive indexes for all years and all states were created by commercial enterprises. These indexes have been supplanted by online indexes that enable nationwide searches with the click of a mouse. Comprehensive online indexes are most often subscription services.

Indexes of a specialized nature are often prepared and posted by volunteers, and are available free of charge. Check cyndislist.com to see if there are online special-interest indexes for any of the places you are researching. Also, check USGenWeb at usgenweb.org for states and counties within your research focus to see if any of their indexes have been posted online.

No matter how good the index, you're likely to come across some mistakes. If you don't find your family in a particular index, don't assume they aren't in the census records. The enumerator may have missed or alphabetized incorrectly during the indexing process, or you may be looking in the wrong geographic location.

Locating a Copy of the Census Microfilm

After you have your list of individuals with their states and counties of residence, their names Soundexed for 1880–1930, and a list of censuses in which you expect to find them, you need to know where you can find the actual census records. Remember, up to this point you've been using only an index.

Microfilmed copies of the federal census records are available in many locations. The National Archives and its regional branches have full microfilm sets that span from 1790 to 1930 (and 1940 when released), with many (though not all) existing indexes. Large libraries and repositories often have complete sets. The Family History Library has a complete collection, and you may borrow and view their copies at their FamilySearch Centers. Images of all the federal census records are online at various subscription services. The most used are ancestry.com and heritagequestonline.com (the latter is available through libraries and other repositories). Footnote.com also has selected census records online.

Is It True?

Census information needs to be corroborated with other records for reasons that I'll explain in Chapter 11. Don't accept census information as completely accurate; use it as a guide.

You'll likely return to census records again and again as you discover new individuals, or as new information calls for a reevaluation of earlier work. Each time you discover a new surname, conduct a complete census search for that family.

LINEAGE LESSONS

Even when information in the census records is accurate, it may be misleading. In the 1870 census record, Lewis Dunn's birthplace is listed as West Virginia, but the 1860 census record shows Virginia as his birthplace. Which is correct? As it turns out, both are correct. The actual place of his birth is in an area that was once a part of Virginia, but it became part of West Virginia when that state was admitted into the Union in 1863. One good source of boundary changes is at http://publications.newberry.org/ahcbp.

The Least You Need to Know

- List and Soundex the surnames you're researching.
- Determine the probable states and counties where your ancestors likely lived.
- Start your census research with the most recently available census records, where you can expect to find the families you're researching.
- Census records are available on microfilm and online.

Making Sense of the Census

In This Chapter

- Locating the census records you need
- Dealing with variations in the federal population schedules
- Searching the census records for clues
- Determining death dates and finding Civil War veterans

Your ancestor's family is sitting around the kitchen table at their farmhouse, talking to the census taker. The baby's crying in the background, while two-year old Hannah is chasing the cat under the table. When was little Jake born? Where was Grandmother Anne born? How old is Mattie? I often imagine these scenes, and wonder how long it took the census taker to elicit all the information he needed on the large family of parents, the 12 children, the grandma living with them, and the assorted farm hands and helpers. What the census taker wrote down forms the basis of the federal population schedules—one of the most frequently used sources in genealogy for both beginners and experienced searchers.

Counting All Those People

The federal censuses from 1790 through 1940 occur every 10 years, and the results are available for public perusal. Normally, the original schedules aren't available for study. Although some have been destroyed, all originals dated before 1930 are saved on microfilm. In some cases, these records were scanned and are available for online viewing. Though some states have had significant losses of early census records, the only serious loss for a whole federal census year was the 1890 population lists, destroyed in a 1921 fire in the basement of the government's Department of Commerce building where they had been stored. Only a very small portion survived the blaze and subsequent water damage, and that portion is now available on the National Archives' microfilm publication M407.

> **LINEAGE LESSONS**
>
> Because of privacy laws restricting access to the census records for 72 years after the census was conducted, the 1930 census records weren't released until 2002; the 1940 census records released on 2 April 2012. The information in these records is extremely important and valuable to twentieth-century researchers.

Index Information

Except for a few instances in which the census taker recopied the lists alphabetically, the census lists in early years were organized in order of visitation. This made the schedules time-consuming to use; a researcher had to conduct a page-by-page search in the county's listings to find a particular family. Various commercial companies and individuals remedied this situation by creating indexes and making them available in books, later on CD-ROMs, and now online. Availability on the Internet is largely by subscription. In other instances, some specific locality census records may be accessible at a state or county website, but they're few compared to what's supplied at the subscription websites.

> **PEDIGREE PITFALLS**
>
> Many indexes published in books were entered into a computer and then sorted. Names typed incorrectly were sorted incorrectly. Andrew Williams, mistyped "Willaims," or John Smith, mistyped "Simth," won't appear in their proper sequence in the publication. Most of those now available on Internet subscription websites have been reindexed to eliminate many of the previous errors. In some cases, the reindexing process introduced new errors.

From 1790 through 1840, the head of the household was the only one listed by name; others in the home were categorized by gender and age only. Therefore, the indexer couldn't include the names of all those residing in the dwelling. If your ancestor was a child during the 1790–1840 censuses, you won't find the ancestor by name; you'll need to search for the head of the household in which that child may have lived.

The census records of 1850 and later included all individuals' names living in the dwelling, but most of the early indexers picked up only the head of the household for the index. This has been remedied. For example, Ancestry (ancestry.com), a subscription service, has created all-name indexes. See cyndislist.com under the category "Census" for the availability of other census records from selected locations.

No matter how complete the index, you may need to be creative in searching it. The name of the head of the household may have been mistyped in the preparing of the index in the twentieth century, the indexer may have misread it, or the census taker could have erred while collecting the data. Remember that many inhabitants were unable to read or write. Census takers spelled the names the way they sounded. Taking this approach was sometimes a problem compounded by the heavy accents of immigrants. Check all variations of the name.

The head of the household isn't necessarily the father. It may be a widowed mother, relative, guardian, or other person. Occasionally, there was a separation, in which case you'll find both the father and the mother as heads of separate households, sometimes listed as widower and widowed instead of divorced or separated.

Soundex (Index) Microfilm 1880–1930

You already have the Soundex code for the surname from following the steps outlined in Chapter 10. You need that code to find the indexed listing. The Soundex microfilm is arranged first by state, then by surname code, and then by first name.

Because the Soundex brings together all similar-sounding surnames, there'll be many different surnames listed within that code. After you find the code on the microfilm, you'll find that within the code it's sub-indexed by the given name of the head of the household. That cuts the search time considerably. For example, you code your surname and search the Soundex for that code. Once you find it, you observe that to find the surname with the first name of Thomas, you must roll the film to the first names starting with T and then locate any Thomas(es).

```
NORTH DAKOTA M1580
 1. A-000 Irak—A-536 Bunt O.
 2. A-536 C.—B-200 Isebell
 3. B-200 J.B.—B-356 William
 4. B-360 A.J.—B-531 Thomas L.
 5. B-532 (N.R.)—B-625 William O.
 6. B-626 (N.R.)—B-653 Willie
 7. B-654 Agathe—C-462 Winnie
 8. C-465 Adelade—C-642 Katherine
 9. C-642 Lars—D-400 Iva
10. D-400 J.C.—E-200 Tonnes
```

Here's an image of the first 10 rolls of North Dakota Soundex from the catalog.

What Will the Soundex Cards Show?

There are four different census cards that appear in the various Soundex indexes: the Family Card, the Continuation Card (included when the family information doesn't fit on one card), the Individual Card (for those who lived in a home in which the head of the household bore a different surname), and the Institution Card for those listed by institution name instead of a surname. Indexers extracted and entered certain items from the full listings onto these cards to form the Soundex index. Family Cards and Individual Cards are especially informative to genealogists. If your ancestor, Jonathan Carlson was living with his grandfather, Jackson Martensen, Jonathan would be listed on an Individual Card. That card would include Jackson Martensen's name and the reference numbers of the Martensen household so you could then obtain that listing.

The census records list the head of the household with his or her age, birth place, and citizenship. This information is followed by the city, street, and house number for the person's address. Then those living in the same home are listed, along with their relationship to the head of the household, age, birth place, and their citizenship. Most important to you, though, is the upper-righthand corner of the card, where you'll find the volume, ED, sheet, and line numbers. You need the ED and sheet numbers to find the whole listing. Remember, the Soundex is only an index coded by the sound of the surname; it doesn't contain everything available in the full listing. It's important to copy the references from the Family Card; you'll need them to obtain the full listing on another set of microfilm.

A shortcut is online at ancestry.com, a subscription-based website where there's an index to the 1920 census records making it unnecessary to soundex the surname. Use their search engine for easy access. Also at this site is the every-name index to the 1930 census records making the necessity of soundexing those surnames obsolete, too.

TREE TIPS

Don't confuse the supervisor's district on the census sheet with the Enumeration District (ED) right below it. It's the latter (the ED) you need to find the full listing on the microfilm.

Using the Microfilm Catalog

You used one set of microfilm to access the Soundex and to get the needed reference numbers. Specifically, you should now know from the Soundex the state, county, ED number, and page number for your ancestor's census entry. Go to the microfilm catalog again. This time, check the section in the catalog for the full 1920 census record (not the Soundex). Determine the microfilm publication number and the roll you need. If there's more than one roll for the county, watch for the roll with the ED number you found on the Soundex.

NORTH DAKOTA

1330. Adams Co. (EDs 1-11), Barnes Co. (EDs 1-19), Benson Co. (EDs 20-35), Billings Co. (EDs 12-20), and Bowman Co. (EDs 21-31).

1331. Bottineau Co. (EDs 36-56), Burke Co. (EDs 32-41), Burleigh Co. (EDs 57-69, 267, and 70-76), and Divide Co. (EDs 42-49).

1332. Cass Co. (EDs 1-2, 12-21, 5-11, and 22-26) and Dickey Co. (EDs 77, 257, 78, 260, 79, 259, 80-82, 258, 83, 84, 262, 85, 256, 86, and 87).

1333. Cavalier Co. (EDs 27-42), Dunn Co. (EDs 50-59), Eddy Co. (EDs 88-95, 266, and 96), Emmons Co. (EDs 97-113), and Griggs Co. (EDs 121-128).

1334. Foster Co. (EDs 114-120), Golden Valley Co. (EDs 60-68), Grand Forks Co. (EDs 54-65, 43-53, and 66-72), and Grant Co. (EDs 69-81).

1335. Hettinger Co. (EDs 82-89), Kidder Co. (EDs 129-134, 261, and 135-140), Logan Co. (EDs 155-162, 263, and 268), La Moure Co. (EDs 141-154), and McKenzie Co. (EDs 90-106).

1336. McHenry Co. (EDs 163-178, 264, and 179-182), McIntosh Co. (EDs 183-191), McLean Co. (EDs 107-124), and Mercer Co. (EDs 125-133).

1337. Morton Co. (EDs 134-152), Oliver Co. (EDs 167-172), Mountrail Co. (EDs 153-166), and Nelson Co. (EDs 73-83).

1338. Pembina Co. (EDs 84-101), Pierce Co. (EDs 192-201), Ramsey Co. (EDs 102-117), and Renville Co. (EDs 173-181).

1339. Ramson Co. (EDs 118-128), Stark Co. (EDs 197-210), Richland Co. (EDs 129-148), and Sioux Co. (EDs 182-187).

1340. Rolette Co. (EDs 202-211), Sargent Co. (EDs 149-158), Slope Co. (EDs 188-196), Sheridan Co. (EDs 212-220), Steele Co. (EDs 159-163, 211, and 164-167), and Towner Co. (EDs 168-175, 212, 176, and 177).

1341. Stutsman Co. (EDs 221-226, 269, and 227-243), Traill Co. (EDs 178-189), and Wells Co. (EDs 244-247, 265, 248, and 255).

1342. Walsh Co. (EDs 190-210), Williams Co. (EDs 239-256), and Ward Co. (EDs 211-217).

1343. Ward Co. (EDs 218-238).

This entry appears in the 1920 Federal Population Census catalog.

There They Are!

You now have in hand the roll of microfilm for your ancestor. Finally, you can locate the full family listing. Proceed to the county you need on the film. Look for the ED number you obtained from the Soundex. EDs appear numerically, usually in the upper-righthand corner of the census page just below the supervisor's district. Proceed to the page (sheet) number you found, and look for the listing.

TREE TIPS

Often, there's a stamped page number, as well as one or two handwritten page numbers. Some indexers used the stamped page numbers, while others used the handwritten numbers.

One of your most exciting moments will likely be in locating your first census record. There they are; the whole family—your grandpa, grandma, and all the children, including your mother when she was a child. It's fun to imagine grandpa and grandma answering the census taker's other questions, trying to remember their former homes so they could give the birth place of each child.

After you've found the census listing, hand-copy it in its entirety. Census forms like those shown in Appendix C are useful. Census forms are also available online at genealogy.com/00000061.html and ancestry.com/charts/census.aspx. When viewing the census data, copy the entries exactly. If it shows Jas., don't write James. If it shows William, don't write Wm. If it shows John Charley Harvey, don't write John C. Harvey. If you wish to add a note, add it in brackets so it's clear this was an addition. Be sure to include the ancestors' gender. You may think it's obvious that John is a male and Ann is a female. But what about Willie? Later, you may wonder if that was a daughter named Wilhelmina or a son named William. Nothing is too minor and anything can later be helpful. If there's a microfilm reader or printer, make a micro-print or hand-copy it all carefully. Note the neighbors, too.

TREE TIPS

Even if you know there's an error in the listing, copy it exactly as it appears. Be sure to include everyone listed in the home, such as boarders, servants, and so on. You may later discover a relationship.

Using the Pre-1880 Schedules

Indexes for the censuses from 1790 through 1870 aren't available in Soundex. They've been indexed by a variety of indexers, however, and are available in published books and online. Though the indexes vary, they usually show the name, county, township, district or ward, and a page number for each census entry.

After you have the indexed entry, determine from the catalog which roll you need. For any pre-1880 census record, you won't need to determine an ED number; they're indexed by the page numbers (and townships), which will lead you to the listing. In a few minutes, you'll likely find that long-sought listing.

The First Census: 1790

Marshals were required to list the number of inhabitants within their districts. They were to omit Indians who weren't taxed (those who didn't live within the towns and cities) and list those who were taxed. They listed freepersons (including *indentured servants*) in categories of age and sex. The rest—that is, slaves—were counted as "all others."

DEFINITION

An **indentured servant** is someone who enters into a contract binding himself or herself into the service of another for a specified term, usually in exchange for passage to the country to which they wish to travel. The number of years of servitude varied, but it was usually four to seven years.

Free white males in the 1790 census records were listed by two age groups, those 16 years of age and upward, and those under that age. The total number of free white females were listed without an age distinction. Only the head of the household was listed by name. Consider this example: John Jackson, was age 40 and had a son, age 18; a son, age 12; a wife, age 38; and two daughters, ages 8 and 10. This family would be listed as two males 16 or over, one male under 16, and three females.

In 1908, the federal government transcribed and printed the 1790 census records for all available states: Connecticut, Maine, Maryland, Massachusetts, New Hampshire, New York, North Carolina, Pennsylvania, Rhode Island, South Carolina, Vermont, and the reconstructed census of Virginia. The 1790 census records for Delaware, Georgia, Kentucky, New Jersey, Tennessee, and Virginia were lost or destroyed. The printed 1790 census records are available in most large libraries. A reprint edition by the Genealogical Publishing Company, published in 1952, made the set widely available. The census bureau website has the 1790 census online at www.census.gov/prod/www/abs/decennial/1790.html.

TREE TIPS

When the census schedules show age brackets such as males 10 to 16, males 16 to 26, and so on, the ages included within the category are actually one year under the next category. For example, males 10 to 16 includes males through age 15, and males 16 to 26 includes males through the age of 25.

Searching the Microfilmed 1790–1870 Census

Locating the census record you need in a repository that holds the microfilm is a simple process:

1. Determine the state and, if possible, the county.

2. Locate published indexes; find the head of the household and note the state, county, township, and page number listed in the index.

3. Using the information found, check the microfilm catalog for the appropriate film and roll number.

4. Using the page number and township or ward from the index, find the family's listing on the roll of microfilm and copy all data shown.

Creatively Using Sparse Information

Let's say you're tracing Jonathan Calavary, who was born on 3 March 1783, according to a Bible record. You find him in census records in 1830, 1840, and 1850 as head of the household. But who is his father? He was only about seven years old during the 1790 census. He should therefore be listed as a male under 16 in his father's home in 1790. Search the 1790 census records for the state for the name Calavary. You may find a family with a male listed as under the age of 16. With that unusual surname, there won't be many and it will be a starting place for your search.

The 1800 and 1810 Censuses

The 1800 and 1810 censuses were more expansive than previous censuses. The head of the household was listed, and free white males and females were listed by age: under 10, 10 to 16, 16 to 26, 26 to 45, and 45 or older. They also included the number of other free persons in the household (except Indians not taxed), the number of slaves, and the place of residence.

In some early census records, the lists were copied and rearranged alphabetically by the census taker. This loses the advantage of listing the family with neighbors. Most often, however, the lists are in the order the families were contacted by the census taker.

LINEAGE LESSONS

Because only the head of the household is named in the census records of 1790 through 1840, it is impossible to determine which individuals are family members in the household. Some of the others listed by age may not be part of the immediate family. Another relative or a helper may have been living in the home.

The 1820 Census: Males 16 to 18

The 1820 census included the same questions as in 1810. It also added a category for males 16 to 18, while retaining the 16 to 26 category. This census gathered the number of those not naturalized; the number engaged in agriculture, commerce, or manufacturing; the number of "colored" persons; and the number of other persons, with the exception of Indians.

TREE TIPS

In 1820, the males listed in the 16 to 18 column are also included in the 16 to 26 column. Keep this in mind when you're figuring the total number of people living in the household.

1830 and 1840 Censuses Narrow Age

In 1830, the age categories were narrowed, enabling researchers to establish ages with more precision. The categories for males and females were as follows: under 5, 5 to 10, 10 to 15, 15 to 20, 20 to 30, 30 to 40, 40 to 50, 50 to 60, 60 to 70, 70 to 80, 80 to 90, 90 to 100, and over 100. The number of those who were "deaf, dumb, and blind" and the number of aliens were included, too. In addition, the number of slaves and free "colored" persons were included by age categories.

The 1840 census records contained the same columns as 1830, with an addition important to genealogical research. A column was added for the ages of military war pensioners (usually for Revolutionary War service). Also added were columns to count those engaged in agriculture; mining; commerce; manufacturing and trade; navigation of the ocean; navigation of canals, lakes, and rivers; and learned professions and engineers. These census records also included the number of individuals in school, the number in the family over the age of 21 who couldn't read and write, and the number of "insane."

The value of knowing the age of pensioners in the 1840 census is immense. The pensioner may have been the soldier, the widow, or another entitled person. Here are two examples: Mary Conklin at age 97 was living with Mary Montanya in Haverstraw, Rockland County, New York. John Jones of Metal, Franklin County, Pennsylvania, was living at the remarkable age of 110. This listing of pensioners was extracted and published by the federal government in 1841, with a reprint by Southern Book Company in 1954 and subsequent reprints with an added index by the Genealogical Publishing Company.

TREE TIPS

The medical profession wasn't as advanced as it is today. Even individuals afflicted by senility, retardation, and misunderstood behavior were likely listed as "insane."

Everyone Has a Name in 1850

The 1850 census was the first to require the name and age of everyone in the household. Its value to genealogists increased dramatically as a result. A dwelling number and family number was assigned to each listing in the order of visitation. Queries included name, age, sex, "color," occupation, value of real estate, birth place, whether married within the year, whether attended school within the year, and whether they could read or write (if over the age of 20). Additionally, they inquired into whether any were deaf-mute, blind, insane, "idiotic," or a convict.

Census records provide useful clues in naming patterns, and in identifying the household as of a certain date, occupation, and migration. The value of real estate gives some idea of the worth of the family and is an additional clue that there may be land records available in the county.

Those Others Living with the Family

Note carefully those living with the family because they may be relatives—perhaps a mother-in-law or a married sister. Importantly, note also the families enumerated at that address a few listings before and after your family because they, too, may be related. If not, they may still assist the search because families usually didn't move in isolation. When migrating from one area to another, they were often accompanied by family or friends. That information may lead you to another residence.

Let's say you're tracing a family listed in the 1850 census records of Montgomery County, Tennessee. The older members of the family were all born in North Carolina. You know they moved between 1844 and 1846 because a child was born in Tennessee in 1846. The surname is common. Where do you look in North Carolina? Check the neighbors. You notice that some of them were also born in North Carolina and, judging from the birth places of their children, seem to have moved around the same time. Look in the 1840 census index of North Carolina for the surname, and the surnames of the neighbors. Finding similar names grouped together in one North Carolina county may help narrow your search.

LINEAGE LESSONS

The occupations listed in the census records may help you. If there were several by the surname in the county who were cabinetmakers, you may suspect a relationship. If they were farmers, look for land transactions. If your nineteenth-century ancestor was a doctor, perhaps the medical school he attended has data. The census records may even point you to military records when it shows "sailor" or "Col., U.S. Army" as the individual's occupation.

Slaveholders and Slaves

There were separate slave schedules taken during the 1850 census. The name of the slave owner, the number of owned slaves, and the number of former slaves now freed are listed. The slaves, however, were not listed by name. The 1860 census was the last one to gather data on slaves and slaveholders. Slaves were shown only by age, sex, and color, and handicaps such as "deaf & dumb, blind, insane, or idiotic." Slaves over 100 years of age were supposed to be named, but there were few. These schedules have been microfilmed and are available at ancestry.com.

Leads and More Leads

The clues that result from using a census listing are many. Assume your family was listed in 1850 Pennsylvania, and part of the information shown is as follows:

Name	Age	Sex	Birthplace	Occupation	Real Estate
SMITH, Jonathan	39	m	Virginia	Farmer	$100
Mary	37	f	Virginia	keeping house	
Barnabus	11	m	Virginia		
Catharine	9	f	Maryland		
Joseph	7	m	Pennsylvania		
Martha	3	f	Pennsylvania		
Jessie	1	f	Pennsylvania		
JORDAN, Barnabus	65	m	Virginia	retired	

How does all this information help? You now know Jonathan Smith was born circa (ca.; about) 1811 in Virginia, and his wife was born ca. 1813 in Virginia. Their son Barnabus was born ca. 1839 in Virginia. Catharine's birth place indicates the family moved to Maryland between approximately 1839 and 1841. Soon thereafter, they moved to Pennsylvania, in time for the birth of Joseph ca. 1843, followed by Martha and Jessie. You now have an idea of this family's migratory pattern. In addition, because Jonathan Smith was a farmer and had real estate valued at $100.00, you can check for a deed to the farm. Even knowing the township is helpful; using that, you can consult county histories to search for early inhabitants and examine newspapers for community columns.

LINEAGE LESSONS

After using the census records to establish an approximate birth for a daughter who was eight years old in 1850, you can record that she was born "ca. 1842." Don't record the birth year without including "ca." (meaning *circa,* or about) because the year may not be correct. In 1850, the census year was measured from 1 June 1849, through 31 May 1850. The actual birth month affects the age calculation. Aside from that, there are many errors in ages, caused sometimes by informants who didn't have accurate information.

What about Barnabus Jordan, aged 65, living with the family? You note that he was born in Virginia (as was Mary Smith, Jonathan's probable wife), and that Jonathan and Mary Smith named a son Barnabus. Barnabus Jordan may be Mary's father. Though this census information doesn't supply the required proof, it provides another lead to follow.

Do you stop here? No. Check the listings before and after Jonathan Smith. Two listings before is a Martin Smith who is 69 years old and was born in Virginia, and Mary Smith who is age 67 and was born in Virginia. Could they be Jonathan's parents? Three listings after Jonathan is Thomas Gordon, 40, and Matilda Gordon, 35; both were born in Virginia. Among their children is a son, Barnabus Gordon. Perhaps Matilda is Mary's sister. The name similarity of the child Barnabus, together with the ages of Mary, 37, and Matilda, 35, who were both born in Virginia, would suggest they may be sisters. Although you will need to investigate further, you have another clue.

TREE TIPS

Use the address provided by the 1880 census records to examine the city directories for a few years before and after the 1880 census. You may find relatives living with or near the family during some of those years.

The 1860 and 1870 Censuses

The 1860 census records contain information similar to what is in the 1850 census records, but it includes a column for the value of personal property and a column for paupers. There were also separate slave schedules, as in 1850.

In 1870, two columns were added titled "Father Foreign Born" and "Mother Foreign Born." (If marked, this may lead to a search of naturalization records and ship passenger lists.) Also, if an individual was born or married within the year, the month of the event was recorded.

1880 Censuses

The 1880 census records include the birth place of the father and the mother (state and country only) to the population schedules. This may prove an important resource, but use it with caution. Often, the informant didn't know where his or her father or mother was born, so he or she guessed. The 1880 census records provide another entry important in your search: the relationship of each person listed to the head of the household, such as "wife," "brother," "mother-in-law," "boarder," or other. Whether they were single, married, widowed, or divorced was noted. Another helpful addition is the address of those who lived in cities or urban areas.

Special 1885 Federal Census

Five states and territories conducted an 1885 census funded partly by the federal government: Colorado, Florida, Nebraska, the Dakota territory, and the New Mexico territory. These records are at the National Archives in Washington, D.C.; only those for Colorado and Nebraska are on microfilm.

Special 1890 Civil War Census

With a few minor exceptions, most of the 1890 federal population census records were destroyed by fire. A special census was taken that year for Union soldiers, however, and its records were destroyed only partially. It's missing states and territories from A through Kansas and part of Kentucky, but the rest survives and includes the soldier's rank, company, regiment or vessel, enlistment, discharge, length of service, post office address, and some other remarks. This is available on microfilm in the National Archives.

Searching the 1880–1930 Census Using Microfilm

Let's consider the search path to find a listing in the 1880 through the 1930 census records. You will use Soundex to identify the codes:

Step 1:

- Code the surname (see Chapter 10).

- Determine at least the state and, if possible, the county.

- Use a catalog to determine the appropriate Soundex microfilm and roll number depending on the coded name.

- Find the entry on the Soundex microfilm.

- Copy all the data shown, particularly the county, the enumeration district (ED), and the page (sheet) number.

Step 2:

- With the numbers you found on the Soundex, return to the microfilm catalog. Look for the state and county (and ED if the county is on more than one roll) to find the proper roll on which to obtain the full listing.

- Using the microfilm roll, find the family's full listing on the microfilm.

- Copy all the data shown.

Changes in the 1900, 1910, 1920, and 1930 Censuses

The 1900 census records have the month and year of birth in addition to age and the number of years married. The mother was required to list how many children she had borne and how many were still living. For immigrants, the data includes the year of entry, the number of years in the United States, and whether he or she was naturalized. The census records also include various other listings.

The 1910 census underwent a number of refinements. One of the most important additions was the question of whether the males were Union or Confederate veterans. This information may lead you to military records.

LINEAGE LESSONS

A serious deficiency exists in locating families in 1910 microfilm because only 21 states have a Soundex or Miracode. For states that don't have a code, you must do a page-by-page search in the counties of interest, or use a subscription service to utilize their indexes. If you can establish the physical address of the family, you can shorten your search by using the National Archives micro-publication T1224 to establish the ED.

In addition to names, ages, and so on, the 1920 census records include the year of immigration, whether the individual was naturalized, and the year of naturalization, as well as other important information.

In 1930, the census included some interesting questions, such as asking whether individuals owned a radio set. Besides personal questions such as sex, race, age, marital status, and occupation, they were asked if they were veterans, and if so, of which war. If an immigrant, the enumerator inquired as to year of immigration or naturalization, or whether an alien. All these details are important in weaving the tapestry of our ancestor's lives.

> **TREE TIPS**
>
> Some nonpopulation schedules are available for agriculture (where you may learn the amount of a crop your ancestor produced in the preceding year), manufacturing, and so on. Read a good discussion of those in the chapter "Research in Census Records," by Loretto Dennis Szucs, in Ancestry's *The Source*.

For the 1930 census records there's Soundex available for only 10 states and partial indexes for 2 states. Complete indexes exist for Alabama, Arkansas, Florida, Georgia, Louisiana, Mississippi, North Carolina, South Carolina, Tennessee, and Virginia. A few counties are available for Kentucky and West Virginia. Under these circumstances, it's best to use the online census records. Otherwise you'll need to access NARA's (National Archives and Records Administration) microfilm publication M1930 to pinpoint the Enumeration District.

The Long-Awaited 1940 Census

For a listing of the questions asked during the 1940 census, go to 1930census.com/1940_census_questions.php. NARA suggests researchers prepare for using these census records by first making a list of all the people they want to look up, and then collecting addresses for these people from city directories or other sources. They suggest you next identify the Enumeration District for each address by using one of the following:

- NARA microfilm publication T1224, Descriptions of Census Enumeration Districts, 1830–1950.

- NARA microfilm publication A3378, Enumeration District (ED) Maps for the Twelfth through Sixteenth Censuses of the United States, 1900–1940.

- Use the 1930/1940 ED Converter utility found at stevemorse.org. NARA indicates this may be useful if the ancestors have been found in the 1930 census records and it's believed they remained at the same address during the 1940 census.

For a listing of finding aids to prepare for a 1940 census search, go to archives.gov/research/census/1940/finding-aids.html.

Mortality Schedules

Starting in 1850, the census taker made a list of those who died during the previous 12 months (for example, June 1, 1849, to May 31, 1850). This mortality schedule is an important search record. It includes the deceased's name, sex, age, "color," profession (or occupation or trade), and birth place, whether the deceased was widowed, the month in which the death occurred, the cause of death, and the number of days ill. The originals for these lists aren't located centrally. In addition to the 1850 mortality schedules, the 1860, 1870, and 1880 census records included similar lists with a few changes. In 1870, the census records included the birth place of the deceased's parents. Mortality schedules were also gathered during the special 1885 federal census. Greenwood's *Researcher's Guide to American Genealogy* has a listing of mortality schedules that notes the physical location of these important records. Ancestry.com provides the mortality schedule images online, with a search feature that allows researchers to locate specific entries.

LINEAGE LESSONS

In addition to the federal government conducting censuses, a number of states conducted state censuses. The states' census records aren't located centrally. The best source for information on these records is Ann S. Lainhart's *State Census Records,* available in many libraries. It explains the differences between the state and federal censuses, the questions asked, and the location of these records.

Can You Trust These Old Records?

You might be perplexed when you find that the census listings you obtained from 1850, 1860, 1870, and 1880 all conflict. How can that be? The problem is you don't know who supplied the answers. Perhaps the father didn't remember the exact ages of the children. Perhaps the mother was shy and became confused when questioned. Perhaps one of the children supplied the information. Add the possibility that the census taker made an error or that he entered the data incorrectly. Also, some problems exist on individual rolls because of pages skipped when microfilming, or because of the poor quality of the microfilm.

If these weren't problems enough, census workers made two additional handwritten copies of some of the original census returns. The copies were distributed to other agencies depending upon the year of the census. This process multiplied the chances of errors. The person writing the copies may have skipped an entry, erred in reading a name, or made any number of other errors.

LINEAGE LESSONS

For instructions given to the census takers for 1790, 1860, 1870, 1880, 1890, 1900, and some later censuses, go to census.gov/history/www/through_the_decades/ census_instructions. For others, insert the census year and the words *census instructions* in your browser.

Now You've Got Something to Work With

Obtaining census records will lay the foundation for the documentation you will build on your family. By comparing the listings from the various census years, you can reconstruct a fairly accurate list of the family members and the places in which they lived. You can even determine something about their background through questions on naturalization or military service. You may get further insight into their monetary worth, those in the family with afflictions, and other details. Always get all the available census records for your ancestors, not just one or two. Every listing is a potential source of important leads and may help to construct a solid, documented line. For additional reading, consult *Guide to Genealogical Research in the National Archives*, Chapter 1.

You'll find that there are many websites that either have transcriptions of all the U.S. Federal Census, or most of them. These include ancestry.com, fold3.com (both subscription based), FamilySearch.org (free) and HeritageQuest (through a library's subscription). Many county websites have transcriptions for their own county. There are still some who either do not use computers, or who believe there could have been an error in an online index, who want to use the microfilm to access the images.

The Least You Need to Know

- In the censuses from 1790 through 1840, only the head of the household was listed by name.
- From 1850 on, the census names each person in the household.
- In 1880, the census records list the birth place (state or country) of the parents, and the relationships to the head of the household.
- Though anyone can use the census records microfilm housed in major libraries and National Archives branches, there are also subscription services online that offer digitized images.
- Census records have many errors created by faulty memories or by the lack of knowledge of the person who supplied the information.

Researching African American Families

In This Chapter

- Identifying ways historical records refer to people of African descent
- Using records from the Civil War and Reconstruction eras
- Researching antebellum free people of color in the North and South
- Identifying the slave owner of enslaved ancestors
- Finding available records for researching enslaved African Americans

Various television programs have explored the family histories of celebrities Oprah Winfrey, Morgan Freeman, Chris Rock, Emmitt Smith, Spike Lee, and Lionel Ritchie. If you viewed these episodes, you may be familiar with some of the issues confronting African American genealogists. The unique history of those of African descent greatly affects how you research your family.

Researching African American families requires all of the same skills discussed in the rest of this book. Talk about the family history with relatives, including parents, grandparents, aunts, uncles, and cousins, and then look for records in your own home. Then you'll be ready to get started with other records.

Two important factors set African American genealogy apart from that of other cultures and ethnic groups: segregation and slavery. Both stem from the racism that African Americans have faced historically in the United States. Both of these factors will affect the way that you need to research your African American ancestors.

Race In Records

Many of the records that you'll use will specify the race of the person. Some of the terms that appear in records are racially offensive. Among the terms used were:

- Black
- Colored
- Negro
- Negress
- Boy (or Girl)
- Nigger
- Mulatto
- Yellow

- Red
- Griffe
- Afro-American
- African American
- African
- Angolan
- Guinea
- Slave

You may come across other terms in your research.

These racial terms were often applied subjectively. A person may be called "black" in one record, but "mulatto" in another. Occasionally, you may also come across individuals of partial African descent reported as "white" in some records. We shouldn't place too much significance on one descriptive word appearing opposed to another. Only by gathering all of the available information about each individual can you reach accurate conclusions.

Race will be an issue both in the existence and maintenance of records and in how you need to use them. Racial discrimination is an important part of the history of African Americans as a group as well as individuals and their families.

Segregation

From the late 1800s until the Civil Rights Movement of the 1950s and 1960s, a system of segregation commonly known as "*Jim Crow*" laws existed throughout much of the United States, especially in the South. Under these laws, African Americans suffered legal discrimination based solely upon race, often requiring that African Americans utilize and attend only designated facilities in schools, restaurants, and stores.

DEFINITION

Jim Crow refers to the former practice of segregating society based on race. Beginning in the 1880s, the Southern states enacted laws that legalized segregation.

During this period, it isn't uncommon to find that records were also segregated. Examples of this are the separate "colored" marriage registers created in many Southern states. If you're unable to locate your ancestors in the regular record books during the period of Jim Crow, check whether the county in which you're researching created a separate record book for "blacks" or "colored people."

Using the research techniques and records described throughout the rest of this book, you should be able to trace your family back to 1870 or 1880. African American research diverges from that of other cultures during this time period.

The Reconstruction Era (1865–ca. 1879)

The period from the end of the Civil War in 1865 until the late 1870s is known as the Reconstruction. During this period, much of the federal government was led by a political group called the Radical Republicans. The focus of political action was on the acceptance of the former Confederate states back into the Union, and the transformation of African Americans in the South from slaves into free members of society.

The Freedmen's Bureau

The largest group of records available for researching formerly enslaved people in the Southern states are the records of the field offices of the Bureau of Refugees, Freedmen, and Abandoned Lands. The Freedmen's Bureau, as it's commonly called, was created by the U.S. Congress in March 1865, under the War Department. The Bureau's original purpose was "the supervision and management of all abandoned lands, and the control of all subjects relating to refugees and freedmen from rebel states." Field offices were in charge of most of the daily operations of the Bureau. Among their records are work contracts between individual freed people and white plantation owners, marriage records, lists of freed people living in Bureau work camps, details of abuse against freed people, and many other informative records.

An important website to examine is archives.gov/research/african-americans/freedmens-bureau. See also the Ancestry.com database "U.S., Freedmen Bureau Records of Field Offices, 1865–1878."

LINEAGE LESSONS

You can find many of these same record types among the records of the Freedman's Bureau headquarters, located in Washington, D. C. And the records of the headquarters cover the entire country. The marriage records have been microfilmed by the National Archives and are available on Ancestry.com in the database "Freedmen's Bureau Marriage Records, 1815–1866."

Only a very small portion of the records has been indexed, abstracted, or transcribed. You may need to do some onsite research to find the information you seek.

The Freedman's Bank

The same day the Freedmen's Bureau was formed—on March 3, 1865—the U.S. Congress also incorporated the Freedman's Savings and Trust Company. The purpose of the "Freedman's Bank," as it was known, was to receive on deposit such sums of money as might be offered by or on behalf of those who had been held in slavery in the United States and to invest those proceeds in stocks, bonds, treasury notes, or other securities of the United States.

The Freedman's Bank records are similar to the Freedmen's Bureau records in their usefulness to genealogists. The Registers of Signatures of Depositors in Branches of the Freedman's Savings and Trust Company (1865–1874) are the most accessible of the Bank's records. Images of these records are available online for free at FamilySearch (http://www.familysearch.org)and by subscription at Ancestry.com.

If your ancestor had an account with the Freedman's Bank, the Registers of Signatures may contain several helpful facts. Some entries identify the names of the depositor's spouse, parents, siblings, and children, the names of their former master or mistress, or the plantation on which they lived.

For more information, read "The Freedman's Savings and Trust Company and African American Genealogical Research," by Reginald Washington, available online at http://www.archives.gov/publications/prologue/1997/summer/freedmans-savings-and-trust.html.

LINEAGE LESSONS

Because of the similarity in their names, many researchers mistakenly believe the Freedmen's Bureau and the Freedman's Bank are related. They are unrelated. The Freedman's Bank was a private company incorporated by Congress while the Freedmen's Bureau was part of the War Department within the federal government.

Voter Registration in the South

Before each Confederate state could be welcomed back into the United States, it had to enact a new state constitution to reaffirm its allegiance. At the time of their secession, most of these states passed Confederate constitutions. To write and enact a new state constitution, each state held a state constitutional convention. To form a convention, the population of each of these states had to vote to do so.

The U.S. Congress, aware of this process, passed a law on 23 March 1867, requiring each of the former Confederate states to create a list of eligible registered voters. For the first time, this would include African American men. Generally, these lists were dated between 1867 and 1869, predating even the 1870 census. These voter registration lists only identify men since women could not yet vote. They also didn't provide the exact ages of voters, though you can assume they were at least the voting age of twenty-one years old. The lists do contain particularly useful information not often found in other records. Many include each voter's place of origin, which may be their birth state or the state they lived in previously. The lists also tell you the length of time each man lived within the state, within the county, and within the precinct. If the voter was a former slave, you can often identify the former owner by comparing the migration routes outlined by these details.

LINEAGE LESSONS

The voter registration lists haven't survived for every Southern state. For those that remain, the information can be extremely valuable.

The Southern Claims Commission

In 1871, the U.S. Congress created a board of commissioners of claims to hear the claims of loyal Southern citizens "for stores or supplies taken or furnished during the rebellion for the use of the army of the United States." The Southern Claims Commission provided compensation to loyal citizens for property given to, taken by, or destroyed by the Union Army during the Civil War.

Many African Americans, including former slaves, filed claims with the Commission. Even more served as witnesses in the claims of their relatives or their former owners. The testimony given by these freed people can tell about their lives both during and immediately following slavery.

A step-by-step search procedure for Southern Claims Commission research is available at slcl.org/branches/hq/sc/scc/steps.htm. See the two subscription websites Ancestry.com and Fold3.com for images.

The Civil War and the End of Slavery

The U.S. Civil War, also known as the War Between the States, was waged from 1861 through 1865. At the beginning of the war, African Americans were barred from serving in the U.S. military, but the Union Army started to use fugitive Confederate slaves in various military functions on the Southern fronts by 1862. In July 1862, the U.S. Congress allowed President Lincoln to use freed slaves in the military, and in 1863, the War Department created the Bureau of Colored Troops. At least 178,000 African Americans served in the U.S. Colored Troops (U.S.C.T.) during 1864 and 1865.

LINEAGE LESSONS

Slaves that ran away from their Confederate owners to the Union Army camps were called **contraband** by Union troops, referring to a phrase for property seized from enemy forces: "contraband of war." When runaway slaves joined the Union Army, they were called "contraband soldiers."

Contrabands, Culpepper County, Virginia, 1863
Library of Congress

If your ancestor served during the Civil War, the first thing you need to find is the compiled military service record. For those who received pensions from the federal government for service in the U.S.C.T., seek their pension files. See Chapter 18 for more on obtaining these records.

African Americans were barred from service in the Confederate Army as well, but this didn't prevent rebel slave owners in the service from bringing male slaves with them to fight. These slaves, though they may have fought, were never officially recognized as Confederate soldiers during the war. The Confederate Congress didn't pass a law allowing the service of African Americans in the Confederate Army until March 1865, but the War was over less than one month later. It appears unlikely that any African Americans ever officially mustered into the Confederate Army.

Confederates Received Pensions, Too

One of the unique record groups created during the late nineteenth and early twentieth centuries were Confederate pensions. While Union Civil War veterans received pensions from the federal government (see Chapter 18), Confederate veterans only received pensions from the state governments where they resided. These state pension laws were enacted starting in the late 1870s, as the period of Reconstruction came to an end. Though never considered a part of the Confederate Army, African Americans in several states were eligible for pensions for service. Check the archives of the Southern states for pension records, even into the 1920s.

> **TREE TIPS**
>
> Be sure to check for Confederate pension records in the state archives of the state where the veteran resided at the time of his application as well as the state where he lived at the time of his service.

Records from the Civil War and Reconstruction eras, though lasting less than 20 years, document the single greatest change in the lives of African Americans. Almost every African American living during this period felt the change, and most of them appear in these records.

The Antebellum Period

The *antebellum* period provides some historically important background for the events that followed it (the Civil War and the Reconstruction Era). In American history, antebellum is generally used to designate the period before the Civil War, specifically the early- to mid-nineteenth century until 1860.

The Underground Railroad

The Underground Railroad was a loose network of safe houses during the antebellum period, supported by abolitionists in the North and South. Many slaves from the Southern states were able to escape to the North using these safe houses and under the protection of Underground Railroad "conductors." Common destinations for fugitive slaves included Philadelphia, New York, and New England, with many continuing all the way into Canada.

If your ancestors escaped to the North, they probably took measures to hide their identities. The *Fugitive Slave Acts* of 1793 and 1850 required all U.S. citizens to assist in the capture and return of runaway slaves. In hiding their true identities, it wasn't uncommon for them to change their names and provide false birth places to record keepers. This will likely cause problems for researchers trying to find their origins. After all, how do you find someone who took deliberate steps to avoid being found?

You may have knowledge of your ancestor's origin through oral history. Some escaped slaves went on to write or dictate "slave narratives," a popular propaganda tool for abolitionists. In other cases, these ancestors may have gone on to serve in the U.S.C.T. during the Civil War. In those cases, their Union pension application files may contain details about their origins not otherwise available.

Free People of Color Before the Civil War

Free African Americans, or "free people of color," lived in every state during the antebellum period. In most states, they suffered legal discrimination, due to the existence of "*black codes.*"

Several states required free people of color to register with the county in which they wanted to live. These "free negro registers," as they are generally called, sometimes contain records of entire families. These registers have been transcribed in published books and online for such states as Virginia, Indiana, and the District of Columbia. See, for example, the Charles City County, Virginia, register at charlescity.org/fnr.

The American Colonization Society (ACS) was founded in 1816 by abolitionists, following the ideas of a mixed-race Quaker named Paul Cuffee. The ACS provided the transportation of free people of color to Africa, where it established the nation of Liberia in 1821. Between 1816 and 1867, the ACS helped in the migration of over 13,000 African Americans to Liberia. The papers of the ACS are currently held by the Library of Congress in Washington, D.C. You can search many of the records in the database "American Colonization Society" on the subscription site fold3.com (formerly footnote.com). The website *African Americans to Liberia, 1820–1904* (liberianrepatriates.com), also has many searchable record extracts.

Many free people of color were once enslaved. Among the records you'll want to locate are those of *manumission*, commonly recorded among the deed books. Slave owners could free their slaves, either immediately or after a period of service, using a deed of manumission. Other slave owners would manumit their slaves through clauses in their wills. The laws of each state specified who could free their slaves, and by what means this could be done legally.

DEFINITION

Black codes were laws established during the antebellum period that specifically restricted the rights of African Americans. These laws existed in the North and South.

Manumission is to willfully free someone from slavery or bondage, as opposed to emancipation, which is the legal end of slavery.

If your ancestors were free before the Civil War, investigate the laws and records available concerning African Americans during this time period. Though this section outlines some of the more common record groups, there are others available in specific locations.

The End of Slavery in the North

Following the American Revolution, the Northern states abolished slavery one by one, either right away or through various systems of gradual emancipation. If your ancestors lived free in a Northern state during the late-eighteenth and early- to mid-nineteenth centuries, you may be able to find significant information on them.

The gradual emancipation laws enacted by states like Pennsylvania, New York, and New Jersey, for example, required that the births of children born to slave women after certain dates be registered with the state. These records can usually be found either at the county or state archives. For those slaves born before the passage of these laws, you will often find the manumission records by which their owners freed them.

The Northern abolition laws included clauses to prevent freed slaves from becoming a financial burden on the state. Generally, these laws required former slave owners to be financially responsible for the slaves they freed.

Throughout New England, it was common for people who moved into a new town to be "warned out," to prevent vagrants from becoming a financial burden on the town. These warnings were literally court orders to move along in many cases, though potential settlers were not generally forcibly removed from town. By the late eighteenth and early nineteenth centuries, as slaves were emancipated in the New England states, it's common to see migrating freedmen included on lists of "warnings" in towns. The practice of "warning out" generally ended during the early nineteenth century.

Before Searching for Slave Owners

Before the abolition of slavery, enslaved people were considered the legal property of their owners. References to slaves appear almost exclusively in the slave owners' records. For this reason, you must identify slave owners. It's usually best to start with the most recent one. Though there's no surefire way to determine who the last owner was, you may be successful using a variety of techniques.

Even if there's evidence of their parents' names, it's still best to start with the most recent generation born under slavery. This generation probably lived the longest, and would therefore have the most information available as a free person.

First Things First

Before trying to identify the slave owner, do the following:

- Gather *all* available information on your enslaved ancestor, until the time of death.

- Identify all siblings and other family members of your enslaved ancestor, to recreate a family group.

- Identify all neighbors and associates of your enslaved ancestor.

- Identify any migration points in your ancestor's life.

Recognize and gather not only the information contained in the federal census records, but also *all* available information. For example,

- If your ancestor survived into the early twentieth century, his or her death may have been recorded by the state. By the second decade of the twentieth century, nearly all states had vital registration programs. A death certificate may identify the birth date and place, and the names of his or her parents. Though the information may not be completely known or completely accurate, it may provide important clues.

- Your ancestor may have purchased land later in life. Deed records may provide valuable information. In some cases, former slave owners rewarded loyal slaves by selling them tracts of land on which to farm.

- Church was usually very important in the lives of our ancestors. Check the records of any churches established near your ancestor's home during that time.

There are many other sources of information, including those records created during the Reconstruction Era. It isn't uncommon to find the past slave owners of your enslaved ancestor identified by name in these records.

Recreate a Family Group

Normally, we identify people through the use of their last names, but records of slaves rarely include last names. In most cases, all we have in these records are first names, so we rely on other forms of information to separate them from others with the same first name.

Children born into slavery bear the same status as their mothers. If your ancestor was the slave of John Smith, then it's likely John Smith also owned your ancestor's mother and all of his or her brothers and sisters. The most efficient way to identify your ancestor is to identify him or her as part of a family group.

In other words, if your ancestor is named Jim Johnson, then he would likely appear in records as a slave named James, Jim, or Jimmy, without his surname. Trying to find Jim by himself in the records of a slave owner would be like trying to find a needle in a haystack. You may find quite a few slaves named Jim of the same approximate age as your ancestor.

If you discover, however, that Jim had a brother named John and a sister named Rose, then you'll have a much easier time identifying him. Rather than trying to decide which Jim is your ancestor based on limited information, you can instead look for the group of siblings—Jim, John, and Rose.

Identify Neighbors and Associates

Don't underestimate the bonds formed among slaves working on the same plantations. Even if they weren't closely related by blood, slaves often formed groups of extended families. In many cases, groups of unrelated former slave families continued to live near each other throughout their lives and even migrated in groups.

If you can identify other families that consistently lived near your ancestor's families, you may have identified fellow slaves owned by the same owners. This can aid your research in the same way that identifying your ancestor's siblings helps.

Identify Migration Points

Some records of freed persons report their birth places. The federal census enumerations from 1870 through 1930 include this information, and 1880 included the birth place of parents (see Chapter 11). Vital records and the 1867 voter registration lists may also include this information. The birth places in these records may not always agree; as a researcher you have to ascertain the most reliable evidence.

When you see a birth place listed, it indicates your ancestor's family, as well as the slave owner, lived in a specific place at a specific time. If you were successful in identifying your ancestor's siblings, then you can also use their birth places to ascertain where the family lived at different times during their lives.

Identify the Owners of Enslaved Ancestors

Once you've located as much of the information noted above as possible, you're ready to start trying to identify the owners of your enslaved ancestors. There's no one way to accomplish this task. Sometimes you may pick out a likely candidate, discover he wasn't the right person, and have to start over with someone else.

Using the birth places of your enslaved ancestor and his or her family members, try to identify white men in the same area who appear to have followed the same migration path. For example, if your Mississippi ancestor was reportedly born in Georgia in 1830, search the 1850, 1860, and 1870 federal census population schedules for any white people also born in that same time and place.

Once you have a list of men who could have been your ancestor's slave owner, try to determine which one of them (if any) may be the correct one. To do this, locate records that would identify each of their slave holdings, beginning with the 1850 and 1860 federal census slave schedules. The records listed in the next section may also help you identify the slave owner.

Upon further investigation, you should be able to narrow down your list of candidates. You may sometimes eliminate all of your candidates. When this happens, broaden your focus and try again. Eventually, you should be able to identify your ancestors among the slave holdings of a specific slave owner. Once you've accomplished this, continue to research the slaves owned by the family, using the records discussed later in this chapter and any others you're able to locate.

A Note About Surnames

Not all slaves used the surnames of their owners. Some genealogists and historians say roughly 35 percent of all former slaves actually used the same surname as their owners. In some areas of the country, this percentage was in fact much higher, while in other areas, the percentage was much lower. With such uncertainty, the only conclusion you can make is that you shouldn't assume one way or the other. When trying to identify your ancestor's former owner, don't put too much emphasis on people with the same surname as your ancestor, but don't ignore them either.

Records Specifically Concerning Slaves

Slaves appear in very few records on their own terms. Most of the records naming slaves are property records filed under the names of the slave owners. Use caution when using these records. Many of them identify slaves by only their first names and perhaps their approximate ages. In rare instances, you may discover some that provide surnames or family relationships for slaves.

Among the records that commonly name slaves are the following:

- **Probate records** This record group includes wills, estate inventories, and estate distribution records. Because slaves were considered valuable property, slave holders often bequeathed them to their heirs. Even without a will, though, the slave holders' estates had to be appraised for distribution. These inventories and appraisement records often provide the name and age of each slave.

- **Guardianship records** Closely related to probate records, guardianship records account for property inherited by a minor. It isn't uncommon to discover slaves "hired out"; that is, slaves sent to work at another farm or plantation for pay. This extra income helped to support minor heirs, providing money for education, clothing, food, and any other expenses incurred in their support.

- **Bills of sale** Slave holders sometimes sold some or all of their slaves to others. The bills of sale include the names, ages, and sometimes physical descriptions or other personal information about the slaves being sold. You can usually find these records in either the deed books or separate chattel books at the county courthouse. It's also quite common to find mortgages on slaves if the owner had significant debt.

- **Tax assessment records** As previously noted, slaves were considered extremely valuable property. Beginning early in the colonial period, governments charged taxes for their ownership. Many of the tax lists simply count how many slaves each slave holder owned. Occasionally, slaves were identified by names and ages on these records.

- **Church records** Many slave owners brought their slaves to church with them. Roman Catholic slave owners in Maryland and Louisiana, especially, were often diligent about having their slaves baptized, married, and buried within the church. You may also find records of slaves in churches of other denominations, including Protestant Episcopal, Methodist (South), and Baptist.

- **Court records** The most common reason one would find records of slaves in court would be if there was a dispute over the division of an estate. You may also find slaves who served as witnesses in criminal or civil cases, or slaves who sued for their own freedom.

LINEAGE LESSONS

The most famous freedom suit involved that of a Missouri slave named Dred Scott in 1846. This case went all the way to the U.S. Supreme Court in 1857. Unfortunately, the Court ruled against Scott, who died in 1858. The case brought attention to issues surrounding slavery, and may have been one of the contributing causes of the Civil War.

• **Runaway slave advertisements** Enslaved people ran away for many reasons, most often to escape slavery or to rejoin family members, including spouses, parents, and children. When slaves ran away, especially in the days after the passage of the Fugitive Slave Act, their owners would post a reward for their "negroes" in the local paper, both where they lived and wherever the slave's distant family members lived. These ads usually contain the first and last name (if any), approximate age, and physical description of the runaway slave. They sometimes contain details on the slave's family, as in the example I've provided.

> **Forty Dollars Reward.**
> Ran away on Saturday morning, a Negro Man named ZACK, calls himself ZACHARIAH DAVIS—formerly belonged to Mr. Robert Lindsey, at Long Green. He is about 5 feet 10 inches high, stout made, a very black smooth handsome face, a scar on the right side of his face below the temple, a remarkable sharp pointed tooth on the right side of his upper jaw, wears his hair plaited, about 21 years of age, is fond of dress, and hesitates much when speaking—took with him a black coat, ruffle shirt, striped waistcoat, white dimity trousers, and an olive coloured great coat fashionably made and nearly new. It is supposed he went away with a free black woman, named Lucy Wallace, who passes for his wife; he will probably change his cloaths, and may have a forged pass.
> Twenty Dollars will be paid if taken in this city or state; and the above reward if taken at any greater distance and delivered to his master.   WM. RABORG.
> ☞ All persons are hereby forwarned from harboring or employing the above described negro man, at their peril.   W. R.
> may 27   dtf

Runaway slave advertisement from Maryland, 1811.

- **Manuscript collections** Many slave owners were diligent in keeping records concerning the management of their plantations. These journals and other papers may contain lists of the births of slave children, records of slave purchases and sales, and even lists of clothing or other supplies provided to slave families. Other interesting contextual information in these documents may detail crops planted and grown in a given season. Family bibles of slave owners can also prove to be rich resources, as these often contain the births and deaths of slaves. Many manuscript collections pertaining to slave owners' records are in the Special Collections sections of historical societies and university libraries. To see if there's a collection of records of your ancestral slave owner, try checking the National Union Catalog of Manuscript Collections (NUCMC) at the Library of Congress website (loc.gov/coll/nucmc).

Another publication, *Records of Ante-Bellum Southern Plantations from the Revolution through the Civil War*, was produced on microfilm by University Publications of America. The full publication contains several individual "Series" organized by location, and over 1,500 reels of microfilm. These microfilms include images of the actual records of antebellum plantation owners.

LINEAGE LESSONS

It may be difficult to locate the *Ante-Bellum Southern Plantations* microfilm publication, but check university libraries. The guides to each series of this publication are available on the LexisNexis website, at lexisnexis.com/academic/upa_cis/default.asp?t=350.

During the antebellum period, slaves were considered extremely valuable property. At one point in the South, a male slave in his early 20s could sell for over $1,000. Because of this high monetary value, slave owners were usually careful to keep thorough records concerning their slave holdings. In many cases, it may be easier to locate information about enslaved people than about their poor white counterparts.

Though many people believe that it's impossible to research the genealogy of African Americans due to slavery, with knowledge of the history and records pertaining to your ancestors, you can have great success.

The Least You Need to Know

- Many records you use will specify race.
- The Reconstruction Era after the Civil War spawned important records, such as the Freedman's Bureau and Freedman's Bank records, the Southern Claims Commission records, and a variety of others.
- Military records for the Union and Confederate Armies can add to our knowledge of African American families.
- To successfully research an African American family, we must identify previous slave owners and re-create a family group.
- A variety of traditional research documents such as probate records, guardianship, bills of sale, and others can provide useful information.

Corresponding Effectively

In This Chapter

- Knowing to whom to write letters and emails
- Encouraging responses to your letters and emails
- Writing clear, concise messages
- Cutting research costs

Writing letters may aid your search substantially. Sooner or later, everyone needs to do it. Contrary to popular belief, not everything is on the Internet!

You may have gotten away from letter writing and rely on the telephone. There are times, however, when a written request is more effective. This is especially true when you're contacting a busy office. If you interrupt a hectic workday, it's too easy for someone to give you a negative answer on the telephone. And many times, short of going to a location personally, a written request will have to do.

Keep in mind your goal and the responder's goal. You want some data. The person responding wants to supply the data with the least amount of effort and time possible. Your letter will be far more effective if you remember certain basics. Clearly state what you want, send it to the right place, and make a good impression so the responder will take it seriously.

The Mechanics of Correspondence

Before we get into what to say in a letter requesting information, let's talk about some basics that should be a part of every letter. Later in this chapter we'll discuss emails, too.

Make It Look Neat

Use standard letter-size paper when writing. It's easier for the recipient to file. Make your letter attractive. Crowding the page leaves the impression that reading it will be a chore. Leave sufficient margins on all sides, and leave a blank line or some space between paragraphs to set them off. Send typed letters whenever possible; if you can't, take care to write legibly, neatly, and briefly.

In this electronic age, many letterheads are prepared on the computer. Use a large, clear font. Those with poor eyesight may mistake the 1 in your letterhead for a 7 if it appears in fancy script or is too small.

TREE TIPS

Keep a copy of the letters you send, at least until you get an answer. It's easy to create a file or document of "outgoing mail" and to add each letter as you write and send it. When you need to refer to a stored letter, use the "find" feature of your word processor, inserting a few keywords you recall (such as the name of the recipient, the address, or the subject).

Include Your Contact Information

Include your name, address, and email address on the letter. Letters can get separated from an envelope. Without an address, the letter you spent so much time composing may never get a response. Position the address at the top of the first page, not at the end of the letter. Small courtesies that lighten the responder's load put you in a more favorable position for a reply.

Invest in a small rubber stamp with your name and address. When you write to someone and include enclosures, stamp the enclosures on the front so the recipient knows at a glance how to reach you. If the recipient photocopies your enclosures and sends them to someone else, that person also needs to know who supplied the data. (Alternately, use small stick-on mailing labels.)

Spell Correctly and Get the Right Zip Code

It's disappointing to have the letter returned in a few weeks because you misspelled the street name or inserted the wrong house number. Recheck the address. If in doubt, use a telephone book or online directory (see Chapter 6). The post office has assigned four additional numbers to zip codes to assist in getting the mail to its destination. Go to usps.gov and click on "Find a Zip Code."

PEDIGREE PITFALLS

Keep it brief and keep it focused! The shorter your letter, the more successful it will be. The key is to be concise. At first you may need to write a rough draft and then rewrite it. If you have a computer, you can create some boilerplate paragraphs you can insert into letters. One might be to a paragraph to introduce yourself, another might be to request a search for an estate, and so on.

An SASE for Reply

Many repositories and individuals won't answer letters that don't include an SASE. While you're at it, make it an LSASE (a long envelope) in case the responder wishes to send you some documents.

What Should You Say?

When you are seeking information about your ancestor, be specific. A "Please send me everything you have" letter, whether to a repository or to a relative, will almost certainly remain unanswered.

To develop the technique of writing an effective letter to a business or repository, consider your goal. What is it you want? Information about the parents? Children? State your desires clearly. You may need to examine different records, depending on your goal. If you are searching for Ezekiel Madison's parents, you may try an obituary, death certificate, birth record, marriage record, and various others. To find Ezekiel Madison's children, his obituary may help, but his death certificate, birth record, or marriage record wouldn't disclose that information. Each request involves different search techniques. Responses will be more on target if the responder understands your goal. Before you write, decide on one or two things you're hoping to find; then state them clearly in your letter.

LINEAGE LESSONS

Do you want to prove two people were married to each other? Or to prove they married in a specific county? There's a difference! A number of records may prove the couple was married: deeds naming them as husband and wife, receipt of a wife's inheritance signed for by her husband, and more. The question as to whether they were married, however, is different from determining in which county they were married. The latter requires you to look at records that may disclose the location: the recorded marriage record, the family Bible, the war pension application, and others. These subtle differences become second nature as you learn about the records and their use in your research.

"Dear Courthouse"

Writing to a courthouse is somewhat different from writing to an individual or a library. In writing to a courthouse, keep your requests separate. If you need a marriage record and a will, write two different letters. This isn't the time to save postage. If those documents are in two separate offices, you'd be dependent upon a clerk to fulfill your order and then transfer your request to another office. A glitch can develop. If, however, you are ordering two of the same type of document, such as two marriage records, include both requests in the same letter; the same clerk will be handling them both. Always limit any request: if you need more than three or four documents, even if they are of the same type, order the additional ones later. Otherwise, a busy clerk will process simpler requests first, setting your more extensive one aside and perhaps forgetting it.

TREE TIPS

Don't know the exact title of the office you want? Write on the envelope: "Probate Office" (for estates) or "Marriage Record Office" or "Deed Office." It will reach the appropriate clerk.

Alternate Spellings

The clerks of the courthouse will not know that the surname you seek can be spelled in a variety of ways. They will look only for the name you request. If you write for the will of John Critchfield, show it in your request letter as "John Critchfield (Scritchfield, Crutchfield)" and any other spellings you have encountered. Otherwise, the clerk may miss the record.

TREE TIPS

Ancestry's *Red Book* describes some of the collections. Many state repositories now have websites that describe their own collections. Also try usgenweb.com for an organized volunteer effort to provide information by state and county. They may also have volunteers to help answer your inquiries.

Write to the Right Place

Say you want to obtain a deed from Groton, Connecticut. Should you send your request to the registrar of deeds addressed at "County Courthouse"? You need a will from Lynchburg, Virginia; do you send the Campbell County Courthouse a letter?

In both of these instances, your letter would have been unsuccessful if you had sent them to the courthouse. Each would have cost you two stamps (for your letter, and for the SASE you enclosed). If you don't get a response you may assume the information isn't available. All the while, the records could be on the shelves of the appropriate record keeper.

What went wrong? Connecticut doesn't keep such records in county courthouses; they have town halls where the deeds are recorded. You wrote to the wrong place. Lynchburg is one of around 30 independent cities in Virginia that have their own courthouse and are not a part of a county. Use a guidebook or an online website to determine the appropriate jurisdiction.

Several books may assist you in determining the correct jurisdiction: Everton's *Handy Book for Genealogists*, *Ancestry's Red Book*, and Marcia Wiswall Lindberg's *Genealogist's Handbook for New England Research*.

In writing to local governments, keep these points in mind:

- In some New England states, the document you want may be in the Town Hall, or in a probate district office (encompassing several towns).

- In Virginia, your sought-after document may be in the courthouse of an independent city. (There are a few other independent cities in the United States, too, such as Baltimore and St. Louis.)

- The county in question may have more than one courthouse.

Locating Other Addresses

To find the addresses of other places to write, use the many directories available in your library; there are directories of funeral homes, newspapers, libraries, and others. *Directories in Print* will widen your eyes with the listings it includes. Also helpful is Elizabeth Bentley's *The Genealogist's Address Book* and Juliana Szucs Smith's *The Ancestry Family Historian's Address Book*. Or go to the website of the county or town; usually the addresses are listed there. Start with usgenweb.org, click on the state, and then click on the county. Use these resources to determine the correct address; you want the letter to get there on the first try.

Sending Emails

It is permissible to use email to request your information if you can locate the email address. Follow these rules:

- State that your request is for genealogical purposes.

- Include a "subject" that is reflective of what you are requesting, such as "request a search of the deed index" or "request photocopy of a will."

- In the body of the message include the name of the person, such as "will of William T. Jones, also known as W. T. Jones and Billy Jones." If there are alternate spellings of the surname, include those.

- State an exact date or a span of time, such as "I would like the deed index searched from 1845 to 1870 for John Allen" or "I request a photocopy of the will of George Fraser who died July 18, 1855, in your county."

- Ask for a quote for obtaining copies.

- Provide your full name, address, and email address. Inclusion of a phone number is optional.

State Registration of Vital Statistics

All states eventually provided for registration of vital statistics on the state level. The beginning dates varied from state to state, but most are in the first part of the twentieth century. Before that time, most counties maintained their own registers, but the starting dates of those also varied. Without knowing the beginning date of state registration, you'll be guessing whether to write to the State Department of Health or to the local government. A handy website for checking on vital records of states is the National Center of Health Statistics at cdc.gov/nchs/howto/w2w/w2welcom.htm. There you can get information on cost, the address of where to order, and other helpful information. You can also check "Where to Write for Vital Records" in Ancestry's *The Source.*

What's on the Shelves?

When writing to a repository, knowing its holdings will increase your success rate. Which of the state repositories has the best newspaper collection for an obituary? Which is more likely to have the state's military muster rolls? Look for published guides, which might be available in your library.

What's It Going to Cost?

It's normally a good practice to inquire about the cost of a document or service in advance of placing an order. If you don't, you may be surprised with a fee higher than you had expected. For instance, some New York counties charge $25 or more for mail-order requests to search a record under 25 years old, and $70 or more for records over 25 years old. Their photocopy charges by mail can be as high as $3-$5 per page.

TREE TIPS

Some repositories accept credit cards. If you're in a hurry for the information, you can call to inquire.

Offering to Pay Promptly

If payment will be required for the record you are writing to request, you have options.

- Enclose a check with the exact fee.
- Offer to pay upon being advised of the charges.
- Send a check for a small amount and ask them to advise you before filling the request if more payment is due.
- If you're ordering a birth or death record, or another record with a standard fee, guidebooks or websites may provide the fee amount. If so, enclose a check or money order for that amount with your request.

If you don't know the fee amount, you can insert a clause, such as, "I will pay the cost immediately upon knowing the charge." This may delay fulfillment of your request because the responder then has two options:

1. To send you the information with the bill.
2. To write to let you know the amount.

Never, never send cash. You may feel it's such a small amount that if it should get lost, it isn't a problem. But this creates an awkward situation; the recipient may feel you will question his or her honesty if the funds get lost.

Sample Letters

Letter writing is a skill you can master. When you see the high success rate from observing a few "rules," it will encourage you to explore what can be accomplished by writing. The following examples of letters demonstrate the technique of keeping requests simple but to the point. Develop a few letters of your own and save them as templates to use when you write future requests.

Trying to Find a Will

In the first example, you're writing to get some information from an estate. You're hoping your great-grandfather left a will. Don't limit yourself to requesting the will. If he died without a will and had sufficient property, there may be an *administration*. Depending upon the reply to the letter shown, you can follow up with a request for additional documents.

DEFINITION

An **administration** is normally an estate in which there was no will. The administrator was appointed by the court to handle the estate. Some counties have a document, called a "Petition for Administration," setting out the names of the heirs and possibly their addresses, depending on the state and the time period. The administrator's bond is more widely available. This was required as security for the performance of administrative duties in handling the estate. Though the genealogical information on it is less than you would find in a petition, it still provides clues and may even provide a relationship between the administrator and the deceased.

[Your name and address]

[Date]

Probate Office
[address here]

Dear Clerk of the Probate Office:

I would like to obtain a photocopy of the will of:

John W. Jorgensen, who died 3 March 1842.

If he did not leave a will, I would like to obtain a photocopy of the petition for administration and the administration bond.

Please let me know the cost for photocopies, and I will send the fee immediately. Enclosed is an SASE.

Thank you for your assistance.

Sincerely,

[Your name here]

As you dig deeper, you'll find there are many additional helpful papers involving estates. For now, writing for a will or administration will get you started. See Chapter 15 for more information about the additional paperwork you may find.

The Obituary

Obituaries are of tremendous value to your search, but obtaining them seems to be a stumbling block. First, determine not only the county, but also—if possible—the town or township within the county in which the family lived. The county may have had more than one newspaper; if so, find the one that covered their community. Then, write to the local library where they lived or the public library of the county seat to see if they have copies or films of the newspapers that they can check.

Suzy Que
1111 Apple Blossom Court
Anywhere, MO 12345-6789

Public Library
[address here]

Dear Librarian:

I would like to obtain a photocopy of the obituary of:

Joseph H. Johnson, who died 14 April 1892, in the town of Sunshine.

[Here include a paragraph as to whether you are enclosing some funds, or "Please let me know the cost and I will remit promptly."]

Thank you for your assistance.

Sincerely,

[Your name here]

Ordering a Vital Record

If you're writing to the county for records created before state registrations were required, try the example shown below. Don't unnecessarily limit the date when requesting a marriage record. Judge a span of time from the birth dates of children but allow enough leeway, perhaps 5 to 10 years.

[Your name and address]

[Date]

Register of Marriages
[Courthouse Address here]

Dear Marriage Registrar:

I would like to order a photocopy of the marriage application (or license) and certificate for:

Joseph Cruse (or Kruse) to Virginia Malley, about 1860–1867.

Please let me know the cost and I will remit promptly. [Or, if you prefer, enclose a small check.]

I enclose a SASE for reply, and thank you for your assistance.

Sincerely,

Dear Cousin ...

If you're writing to a relative for the first time, introduce yourself. Let them know your purpose and what it is you're seeking. Include a family group sheet with the letter, which may pique interest, but don't flood the cousin with data at this point in the contact.

[Your name and address]

[Date]

[His or her name and address]

Dear [His or her name],

I have been working on the family history for two years, and just found out from a relative that you and I are second cousins. We are great-grandchildren of George Milliken and his wife Susan Masters. My family is through their daughter Jane, who married Jesse Cooper. My mother told me she lost track of your family after your parents moved—about 30 years ago. I am delighted to now be in touch.

I was told your grandparents passed away many years ago. Can you tell me when and where they died, and where they are buried?

I have also been hoping to find someone with a photo of our great-grandparents, George and Susan. Do you have one that can be copied or do you know of anyone in the family who does?

It would be a delight to hear from you and to exchange information about our families.

Your cousin,

[Your name here]

TREE TIPS

Keep the Post Office 1-800 number close by, and use it to inquire about rates when you're in doubt. Request the basic leaflet on postal rates from your local post office so you can refer to it as needed. Or go to usps.gov, click "All Products & Services," and then click on "Calculate Postage."

A Penny Saved ...

The cost of letter writing has escalated. Not only do you pay more for postage, but also for paper, envelopes, and supplies. You want to make your letters count.

Invest in a postage scale. The small scales go to one pound, or you can get a two-pound meter for a slightly higher fee. It won't take long for you to recoup your investment. Without a meter, you probably often guess at the weight, adding additional postage just to be sure.

The Least You Need to Know

- Everyone needs to write a letter at some time; it's impossible to do all genealogical research online.
- A concise letter will bring more results.
- Don't combine a variety of different requests in the same letter.
- Make the letter appealing to the eye.
- Address the letter properly to ensure that it gets to the right place.
- Know the proper rates and weight to avoid guessing and to reduce postage costs.
- Respect similar rules for email; be concise, provide your full contact information, and ask for a quote.

In Your Ancestors' Footsteps

Grab your suitcase—we're going on the road! In this part, you'll visit the areas where the family lived. You'll walk the same ground they walked so many years ago. You'll learn to use the courthouse records, to visit the cemetery and glean new clues, and to search the fascinating newspapers while there. You'll also learn of the marvelous military records that await you on the trip to the National Archives or its branches. And you'll weave Internet resources into your trip preparations.

Excitement beckons—start packing! Even if you don't plan a personal trip in the near future, the advice and techniques in this part take you farther down the road in your pursuit of your ancestors.

A Little Traveling Music, Please

In This Chapter

- How to plan a research trip
- What to take
- What to do before you leave
- What to do when you get there

You've read some genealogy guidebooks, researched a number of census records, corresponded with relatives and repositories, and developed a familiarity with the kinds of records that will advance your knowledge of the family. Soon you'll be eager to travel to certain counties for a firsthand look at the records created by and for your ancestors. Some of the most exhilarating experiences in genealogy result from delving into original records (rather than published or filmed ones) and from walking the roads trod by your ancestors.

Decide which counties are likely to produce the most information on the most people. Review the information you've accumulated and narrow down the possibilities for the first trip. Your decision will be influenced by many things: time, distance, and cost of travel; priorities, such as a family reunion or an elderly relative in poor health; a burning curiosity about one ancestor in particular; and other personal issues.

Counties Have Ancestors, Too

Before you go too far in planning your trip, be sure you're headed for the right county. Counties are the political subdivisions of states. Their current boundaries aren't necessarily the ones they've always had. Newer counties were carved out of one or more old ones, or were absorbed by adjacent counties. Disputes and/or new surveys redrew the lines between counties. Why should you care? Because the records of your ancestors are in the counties as they were when they lived there.

Some states have forms of government other than counties. Connecticut is organized around towns, rather than counties. In Virginia, there are numerous independent cities, completely autonomous from the counties they adjoin. Louisiana counties are called parishes. Read up on the political divisions of the state as you start your research there. Skipping lightly over the preparation can lead to many disappointments, not the least of which is to find you're in the wrong county.

Your ancestor may have spent a lifetime on the same piece of land, yet may have resided in two or more counties due to boundary changes. If you don't know about the *parent county*, you may miss all the valuable records your ancestor left there. To know where to look, you must find the dates the counties were organized and the names of the parent counties.

To find parent counties, consult Everton's *Handy Book for Genealogists* or Ancestry's *Red Book*. For boundary changes in census years, see William Thorndale and William Dollarhide's *Map Guide to the U.S. Federal Censuses, 1790–1920*. You may find more information by checking the county's website on USGenWeb (usgenweb.org). Niche software AniMap (goldbug.com) shows every county's boundary change from colonial times, not just from census years. See also the *Atlas of Historical County Boundaries* at http://publications.newberry.org/ahcbp.

While you're zeroing in on the county, also find the *county seat;* that's usually where the county records are kept.

DEFINITION

The **parent county** is the county from which a present-day county was formed.

A **county seat** is the town that is the administrative center for a county. Don't assume it is the largest town in the county. A few counties even have two courthouses.

Don't Pack Lightly

Old admonitions about packing lightly for travel don't apply to genealogy. At the minimum, you need these reference materials to make your research trip effective:

- Pedigree charts

- Family group sheets

- Research calendar

- Your notes

- Listing of county formations and parent counties

- Maps of state and county roads

- Prioritized checklists

- Lists of facilities' addresses, phone numbers, and hours

- Names of possible contacts

TREE TIPS

If you belong to a genealogical society, pack your membership card. Some states require membership in a genealogical society to use certain records.

Also useful are travel guides, such as those from the American Automobile Association (AAA) and state genealogical guides. There are excellent genealogical guides for many states. They usually provide historical background, overviews of the state's court system and laws, and sections covering each county's repositories and records. Read a genealogical or travel guide for background information before your trip to the state.

Pack Those Technology Marvels

Do consider taking your laptop computer, digital camera, and cell phone. Optional is a hand-held scanner. You may feel you don't want the extra gear, but it's advantageous to do your note-taking on the computer as you go rather than waiting until you get home. Also, if you use a genealogy program, you can quickly check names and dates you didn't bring, thinking you'd not need them on this trip. If your travel accommodation provides Internet connection, you can continue your research online after the local facilities shut down for the day.

TREE TIPS

Some courthouses don't allow digital cameras past the guards. Be prepared for that possibility. Usually the guard will hold it until you leave.

Record custodians sometimes allow visitors to take digital photographs of documents too fragile to subject to copiers or scanners. If you have learned to photograph microfilm records, you can save the cost of copies. The instant feedback of photographing cemeteries or landmarks with a digital camera prevents the agony of getting home without the sharp photo-documentation you thought you had.

Prepare a Packet and Checklists

For each surname you're researching on this trip, prepare a packet or notebook that includes the pedigree charts, family group sheets, research calendars, and your notes. Review this material, looking for the gaps in your information, and start some lists of what you want to know.

Put in your packet the checklists covering what you're missing and prioritize the information you need. Are you trying to locate the deeds you think must exist because on the 1860 census your grandmother is listed as having $1,000 worth of real estate? Are you missing a marriage date? Are you trying to prove a death date? Are you missing a wife's maiden name?

PEDIGREE PITFALLS

Never take original documents in your travel packet. You may accidentally lose or destroy them. Instead, take *copies* of anything you think will be helpful.

Decide which missing piece you want to tackle first and what records you need to see. Are you most interested in deeds? Estate records? Marriage records? Perhaps you're trying to locate a hard-to-find family history, or you want to search for obituaries in the local newspaper. It's a rare genealogist who has time to exhaust all the possibilities on the first onsite research trip, so decide ahead of time what is most important to you. Go through the list again and add an "if time allows" list.

Have alternate names or ideas because your original plan may be thwarted when you get to the courthouse. At a courthouse to search the 1870 tax records stored in the attic and inaccessible without an escort, I arrived to learn that the escort's mother had died the previous night. The office was shorthanded due to vacations and no one else could help. The trip would have been wasted if I hadn't had a secondary set of objectives.

Map Out Your Strategy

Become familiar with the county you're interested in and its surrounding counties by studying several kinds of maps before your trip. You need present-day highway maps to find your way around and to give you an overview of the local scene. But even if your ancestor's town still exists on present-day maps, try to find maps contemporary with his or her life.

The *Atlas and Gazetteer* Series produced by DeLorme for all 50 states combines detailed maps with lists of historic sites and museums, natural features, scenic drives, and wineries—whatever is pertinent to that state. They list the covered bridges, ferries, and lighthouses; the scenic-drives section features Amish sites and heritage tours, or suggests traveling the Old National Road. Look beyond the road map for historical perspective on your family. You can order DeLorme maps from their website: delorme.com. You can also experiment with GoogleMaps.com, MapQuest.com, and Topo.com.

Try to determine where your ancestors lived in relation to the county seat. Then remember the conditions they faced in trying to get to the courthouse to conduct official business such as recording a deed or getting a marriage license. If they had to navigate some rugged terrain, they may have opted to delay recording the deed. If they lived near the border of another county or state, they may have records at the other location. Understanding the challenges and limitations your ancestors faced may be crucial to your finding the records. Mystified as to why there was no marriage record for a couple I was sure must have married in a particular county, I expressed my puzzlement to the clerk. I learned that, at that time, couples sometimes took the train 18 miles to a town in the next state to marry because there was no waiting period and the age limit was lower.

The Lay of the Land

Topographic maps are another helpful aid to secure before you travel. The detail on these maps may literally take you in a new direction. Farm roads, cemeteries on private lands, and churches are all usually marked on these maps prepared by the U.S. Geological Survey. Done on such large scales that it often takes several maps to cover one county, they're more useful for research purposes than the usual county maps of today. Seeing a now-isolated family cemetery once accessible by a farm road can lead to a sought-after burial site.

> **DEFINITION**
>
> A **topographical map** is a detailed, precise description of a place or region. It graphically represents the surface features, such as elevations and creeks.

You'll enjoy poring over the maps, locating the creeks and ridges, and comparing the features with the deeds you find at the courthouse to determine the location of your ancestor's land. When the deed reads "under and on the great mountains on the branches of Rocky Creek," you'll know where to look.

The maps, also called quadrangle maps, are available for a nominal fee from the U.S. Geological Survey at http://store.usgs.gov. First order the index for the state you're researching. Then use the listings to order the maps you need.

In the following section of a topographic map of the Baldwinsville Quadrangle, New York, note the cemeteries shown.

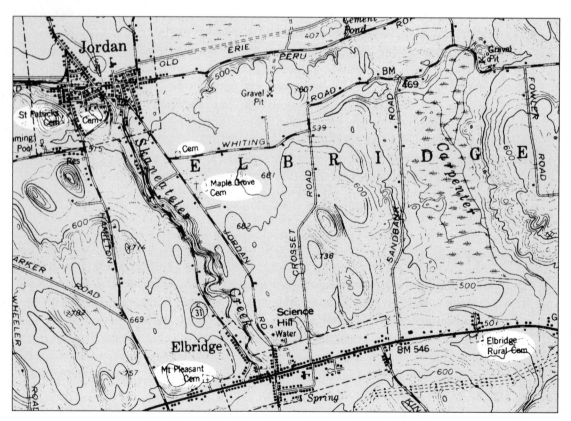

Here's a section from a topographic map in New York.

TREE TIPS

Always seek permission before venturing onto private land. A simple explanation of your purpose will usually get you access.

Have Contingency Plans

Another map for your travel packet is a simple outline map of the state with only the counties and their county seats marked, such as the maps found in Ancestry's *Red Book* or in the computer program AniMap. You may need to refer to this map for ideas of other counties where you might find your ancestor's records. If you're researching in Noble County, Indiana, and you find a deed that says your ancestor was "of LaGrange County, Indiana," you'll want to look at your map to see where that is and consider going there on this trip.

If your research trip is in a county bordering another state, be sure to take a map of the nearby state. You may uncover leads that take you into that state, and you don't want to spend time hunting down a map. With Internet connections commonly available at libraries nationwide, you can probably check for information on research facilities at that next destination and soon be on your way.

TREE TIPS

Take rolls of quarters and dimes to use for parking meters, vending machines, and copy machines. Large facilities often use copy cards, but usually maintain one machine that takes small change from individuals not wishing to purchase a card.

Call Ahead

I can't overemphasize the importance of calling ahead before making a visit. Make a list of all the places in the county you want to visit; you can get ideas from state guides, comprehensive guides dealing with all states, and directories listing museums and historical societies. Try to determine what special collections of materials may be in the county. Call each site you'd like to visit and ask their hours and whether the records are open to researchers during those hours. Be sure to ask if there are holidays, special events, or unusual circumstances that will interfere with your access.

Even if the county or historical society has a website with hours and accessibility, confirm by phone. The website may be outdated or it may not cover irregular events.

TREE TIPS

Never assume you'll be able to have immediate access to public records. Hours and rules for access change. Always call to check on accessibility before making a special trip to a distant repository.

Imagine my surprise at finding a handwritten sign on the courthouse door saying, "Closed Monday for Deer Day Holiday." This rural county closed government offices and schools on the first Monday of the hunting season! Another time, I called ahead but didn't ask the right questions. The courthouse was open on Columbus Day, but I didn't ask if I would be able to research that day. Unfortunately, the small room housing the old records I needed was closed for research because county officials were counting absentee ballots in there. Another time, records I particularly wanted to see were inaccessible because they were being microfilmed.

Pack for Research

Pack comfortable shoes. You may have to do much of your research while standing. Old courthouses have limited research space and high counters. You may have to climb a ladder to reach the earliest volumes stored near the ceiling. Clothing should be "business casual." You'll have better service if you aren't in a sweatshirt and jeans. Take clothes that won't show dirt; old records are dusty and stored in areas that are rarely cleaned.

Now You've Arrived

You always have a tight schedule on a research trip, so use your checklists and the information in Chapters 9, 15, and 16 to make a tentative work plan for each day. Courthouses usually open and close early, but libraries often have evening hours. Small museums and historical societies may have very limited hours; to visit them, you have to plan around their schedules.

Most research facilities are closed on Sundays. Use that day to visit the cemeteries, attend the church services at your ancestors' churches, find the old home place of your ancestors, or visit with distant relatives. Reflect on your feelings as you gaze at the same mountains your ancestors saw, walk the creek bank where they fished, or sit in the church pews they once occupied.

TREE TIPS

Remember that the details of your family history usually interest only you and your relatives. You could mention the purpose of your trip to the attendant and ask if there's anything in their collection you've overlooked.

Engage the People

Talk to the individuals providing services for you: staff members at the courthouse, the libraries, and the Chamber of Commerce; the volunteers at the museum; and tour guides. They may know of a source, record, or individuals you wouldn't find on your own. In one courthouse, a clerk produced an interim report of the survey of the county's historic properties. This working document with its detailed maps, photographs, and historic background of the county's communities was an outstanding research aid, one that I wouldn't have found in the normal course of quiet research.

Old-Timers Can Tell It All

Often, there are individuals in town who knew your ancestors. They were their neighbors or their fathers were in business with them. Their mothers were in the same church groups. These old-timers are delighted to reminisce with someone who hasn't heard their stories dozens of times. They may also tell you things the family won't: "Your grandpa was quite handsome, and he got around some."

The old-timers may have photographs of your family. Ask if you may get them copied while you're in town. Or, if you have a digital camera, photograph their photos on the spot.

Small towns and rural communities sometimes have unofficial historians—those who seem to know all the old stories about their areas. These individuals are often eager to show you where the old tavern stood or tell you that your aunts and uncles and other children from the farms rode in a covered wagon into town for school.

TREE TIPS

Check the telephone books for surnames you're researching. There may be descendants still living in the area. If you can't contact them on this trip, you may be able to reach them when you return home.

While You're in Town

Buy the local newspaper. You may get an idea for another source of information from something you see in the paper. Perhaps there's a farm auction or an estate sale; old books, such as county histories and old photographs, often turn up at these sales. It may be worth your while to inquire.

Check the bulletin boards at the library. Often, local genealogical societies don't have offices; to reach them you must know the officers. Meeting announcements or flyers about their services are often posted in public places. Even if you don't contact them now, you may want to engage someone from the society when you get home to follow up on something.

Pick up brochures on the local historic sites. The information in them will add flavor to your family history, making your ancestors more "real" to you.

Find Religion

If you've determined your ancestor's religion, try to locate the meeting place. Your ancestors usually didn't venture very far for religious services. Check city directories and old maps to help locate places of worship closest to your ancestor's residence.

Investigate the possibility that the religion is still active in the local area. Call the church or synagogue office to locate the old records. There may be membership lists, participants in ceremonies, or a history mentioning your ancestor as a founding member. More often, the structure is gone and the archives, if they survive, have been transferred to the library or a central repository.

Be a Tourist

Absorb the atmosphere. Read the historic plaques. Buildings your ancestors saw every day will help you visualize their times. When was the courthouse built? Your ancestors may have walked up these very steps to get their marriage license. Take a walking tour of the town. Look for monuments inscribed with names; your ancestors' names may be on a war memorial.

Does your ancestor's house still stand? Perhaps you'll have time to drive by it. Check first to see if the neighborhood is a safe one, as neighborhoods change over time. Use caution if you want to photograph the house. The current residents may not understand a stranger's intent when they see a camera pointed at their home.

Drive the country roads. Imagine the days of times long gone, preserved in such serene scenes as the one in the following figure. Visit the old home place if it still exists. There may be a McDonald's on the spot now, but let your imagination replace that with the pictures you saw at the museum. Visit the cemeteries. Information and ambiance await you there. Chapter 16 details how to make the most of this part of your trip.

Before taking off on this trip, immerse yourself in the names of your ancestors and their associates. You want the names of associates in your subconscious so they jump out at you if they're mentioned in a document or monument—such references may lead you to your ancestors. You don't know what you'll find when you get to the county, so anything you have tucked away in a corner of your mind may be useful.

Here's a wonderful scene from History of Oneida County, New York.

The Least You Need to Know

- Advance planning is the key to making the most of your time and money on a research trip.
- Maps are some of the best investments for your research trip.
- List what you want to know, and learn where the answers are likely to be.
- Call ahead to check the hours of the places you want to visit.

Courthouses: Gateway to the Past

In This Chapter

- Using the indexes to find documents
- Knowing what to look for in the documents
- Letting documents tell the tale
- Finding courthouse records on the Internet

I always feel a thrill when entering a courthouse. It never diminishes. There's the excited expectation of what may be on the shelves or stored in a dusty attic. Walking up the stairs, worn with the steps of those who have entered for decades, I am swept with a sense of history. Visions of those family members who may have climbed the same stairs as they came to pay their taxes or to settle their grandfather's estate come to mind. I wonder if I'll find a forgotten record stuffed in one of the metal boxes in the clerk's office that will solve a longstanding genealogical problem. In this chapter, I'll pass on tips to help you find *your* history under all that dust.

What to Expect

Some courthouses have made an effort to carefully preserve their records. They have flattened the original papers and now store them in file folders. Some have even laminated the registers. Others, due to lack of interest or lack of funds, have done little to save their documents. The records may be in deplorable condition. Be prepared for anything.

Another problem is the burned courthouse. Many courthouses were destroyed during the Civil War, or the old structures caught fire, perhaps from an overheated potbellied stove used for heat. Some were set afire on purpose. The clerk tells you, "The original courthouse burned in 'the war.'" Don't become discouraged. I've found that a few additional questions may have surprising results. The first is, "Which records were destroyed?" followed by, "Which were saved, and where are they stored?" It's amazing how often certain records were saved, but the clerk may not offer that information. Ask the right questions!

In this photograph, typical courthouse shelves, with heavy books and metal boxes.

PEDIGREE PITFALLS

Women should wear flat shoes. Men should not wear slick-soled shoes. Floors are often uneven, and stairs can be narrow. Sometimes you'll even need to climb a ladder. Many early records are copied from the originals into large, heavy, bound books. If you're unable to climb stairs or handle the books, bring someone who can.

The History of the Farm

Be sure to take with you the names of the people you're seeking, the records you hope to find, and an idea of the time period. Perhaps at the top of your priority list you noted your desire to examine the deed to the old family farm. If so, your first step is to locate the office that holds the land records. There are a variety of offices in the courthouse: the County Clerk's office, Circuit Court Clerk's office, County Recorder's office, and others. Check the directory that's usually posted on the wall near the front door to find the right office.

Deeds have almost always been indexed. Land was important, so a way to keep track of who owned it was essential. A deed was signed by the seller, who gave it to the buyer as proof of the sale. The buyer then took it to the courthouse. The seller or the witnesses came into court to verify that the grantor signed it. This created a notation in the court minutes (a book recording a day-by-day account of what happened in court) that the court acknowledged the exchange, and the court ordered the transfer of property to be recorded. Usually, the original deed remained with the clerk of the court for several days or even weeks, until he had a chance to transcribe it into one of the deed books. The original deed was then returned to the new owner.

LINEAGE LESSONS

In some colonies or states it was called an acknowledgement if the seller signed the deed personally, but was referred to as proved if the witnesses came into court to verify the signatures.

The Approach to the Clerk

When you enter the deed office, a clerk will offer to assist you. Don't go into the details of your family's history. The clerk is no doubt busy with the day's current activities and has no time to get involved. Instead, because you know what you want, simply say, "I would like to see the deed indexes for 1800 through 1875" (or whatever records and time period you're seeking). Normally, the clerk will take you into the vault and show you where to find the indexes, or just point you to the vault and let you proceed on your own.

Indexes Tripping You Up?

Enter the record room and look around for the index books. Deeds are always indexed under both seller and buyer. There may be separate indexes: one for grantor (seller), sometimes called a Direct index, and one for grantee (buyer), sometimes called an Indirect index. Alternatively, the grantors and grantees may be listed together in one book called a General Index. If so, the grantors may be in the front part of the book, and the grantees in the back. In other variations the left page may be the grantor index, and the right page the grantee index. Or they may simply be intermingled, entered as they were recorded. In this case an additional column shows "to" or "from," indicating whether it was a grantor or grantee. Each index book covers a period of time. Index Book No. 1 may be for 1802–1840, Index Book No. 2 for 1841–1890, and so forth.

For example, if the clerk transcribed the original deed into Deed Book A, page 121, he would then index the deed either in a separate index book under the name of the seller and under the buyer, or in an index created on the first few pages of the original deed book. He would also show the book and page reference in the index. If there was more than one seller or more than one buyer (perhaps the property was owned jointly by John Mathews and his brother-in-law David Donaldson, and sold to George Graham and his wife Martha Graham), the deed was then indexed in each of the sellers' and buyers' names (although there are exceptions to this).

> **LINEAGE LESSONS**
>
> A court clerk may suggest you need only look at the grantee index, and not the grantor index. Don't listen! There can be a variety of reasons why the purchase was not entered in the grantee index, although later the sale appears in the grantor index. The first deed may have been a patent or grant from the state or federal government, inherited, overlooked when the index was prepared, or just not recorded. Always check both indexes.

Sometimes the record was indexed only under the name of the first grantor (seller) or the first grantee (buyer) listed on the document, followed by the notation *et al.* (*et alii*, meaning "and others"), *et ux.* ("and wife"), or *et vir* ("and husband"). If it shows only "John Mathews, *et al.*," "George Graham, *et ux.*," or "Mary Williams, *et vir*," be sure to follow through and locate the actual deed. It will reveal the names of the rest of those involved (referred to as the *parties* in legal documents). Any record in which there are multiple parties should be examined carefully; often it is a transaction between family members.

Making Sense of the Index

Open the index book, and see if it's a standard index you can understand easily. There should be a diagram or chart and an explanation of the indexing system for that county. Read it carefully. If you're still unsure after a few minutes of study, ask the clerk to explain the system.

In the following figure, using the so-called Russell system of indexing, go to the index book with the initial letter of the surname you are tracing. Then determine the first key letter as listed at the top of the figure (*l*, *m*, *n*, *r*, or *t*), to find where in the deed books the surname is indexed. For example, if the surname is Martin, then you're searching in the *M* index. Ignoring now the initial letter of the surname (you are already in the *M* book), search for the first key letter. The next letter is *a*, not one of the key letters (*l*, *m*, *n*, *r*, and *t*), so ignore it. The next letter, *r*, in Martin, is the first key letter of that name. Looking at the columns, you note that Martin is therefore indexed in sections 14, 24, 34, 44, 54, and 64.

In the left column of the same figure, the letters ABCD, EFGHI, and so on, represent the initials of the given name. If you are searching for Abraham Martin, go to section 14 of the index. If you are searching for Mary Martin, proceed to section 44 of the index.

> **TREE TIPS**
>
> Land transfers weren't always recorded. The land may have been inherited rather than transferred by deed, or deeds among family members may not have been taken to the courthouse to save the recording fee. Diligent searching, however, should produce something, perhaps even years after the original purchase, when the land was sold out of the family.

If you're searching a name such as Rowse in the Russell system of indexing, because there are no key letters (*l*, *m*, *n*, *r*, or *t*) following the initial letter of *R*, you would use the Miscellaneous column. (Anything not falling within the key letters is considered miscellaneous.) Abraham Rowse would be in section 16, while Mary Rowse would be in section 46. The Russell index, though prevalent (and still in use), is only one of numerous types of indexes.

TO LOCATE NAMES IN INDEX

Determine first key-letter following initial letter in Family Name. Find section number in the column headed by said key-letter, opposite given name initial desired. Names not containing a key-letter will be located under "Misc." Corporations, etc., will be located under the first key-letter following the initial letter in the first word of the name, or if no key-letter, under "Misc." Always omit the article "The."

Given Name Initials	Key Letters and Section Numbers					
	l	m	n	r	t	Misc.
ABCD	11	12	13	14	15	16
EFGHI	21	22	23	24	25	26
JKL	31	32	33	34	35	36
MNOPQR	41	42	43	44	45	46
STUVWXYZ	51	52	53	54	55	56
Corps., etc.	61	62	63	64	65	66

Here's a sample of the Russell index, arranged by certain key letters.

What Should You Note?

After you determine how to use the index, look for entries that involve your family. The indexed entry will include the name of the grantor and grantee along with the date of the document, date recorded, type of document, book and page where it's recorded, and perhaps a very short property description showing the township, lot number, waterway, or other brief designation.

While you search the index, note the column that designates the type of deed. Typical designations may be warranty deed, deed of trust, gift deed, quit claim deed, power of attorney, partition, or others. Each is a specific way of conveying land or rights. You'll likely become familiar with all these terms and understand the value of each in the course of your research.

Be sure to note the reference given in the index: the book or volume number and the page number. You will need this to find the document. Note, too, the type of document.

LINEAGE LESSONS

When you find the book and page reference, take note of any abbreviations that appear immediately *before* the volume number. There may be separate volumes for specialized records. For example, PA (Power of Attorney), DT (Deed of Trust), or some other designation, may precede the volume number.

Search Strategies in Deeds

While you're working with the deeds, take some time to copy all the index entries for your surnames of interest in the appropriate time period. If, after you return home, you find you need another deed, you can order it by mail, citing the book and page number (or you may be able to order microfilm of the records from the Family History Library).

Be particularly alert to indexed entries that appear to involve several people—they could be family members. The words *et al.* ("and others") can be a tip-off to such transactions. If the document is designated Power of Attorney, Gift Deed, or Partition (involving divisions), or if it mentions an estate, then you should take the time to examine it. These types of documents often show familial connections.

LINEAGE LESSONS

Sometimes the buyer never returned to the courthouse to retrieve the now-recorded document. The original deed of your ancestor's property may be in that old box marked "original deeds" or "unclaimed deeds" on the top shelf, gathering dust.

Is He Augustus W. Redman, A. W. Redman, or Gus?

Consider possible alternate versions of the first or given name, including initials, the first and middle name reversed, and even nicknames. George Washington Smith may appear as George, Washington, G. W., George W., Washington G., W.G., G. Washington, and Wash. It's important to know all the possible variations. If you don't know Patsy was a nickname for Martha, Nabby for Abigail, Jane and Jennie for Virginia, Polly for Mary, and others, you may miss the listing you're seeking. Christine Rose's *Nicknames Past and Present* includes many hundreds of nicknames, all cross-indexed.

It's There Somewhere

You searched the index and noted a deed to what appears to be the old family farm, listed as Book B, page 510. Look around and see if you can find some books that are marked on their spines with the word "Deeds." Watch for Books A, B, C, and so on.

It's permissible to remove a deed book from the shelf and put it on the counter for examination, but be sure to put it back (in the same spot) when you're finished. The counters on which you will be working are usually high and slanted. Other researchers (perhaps title searchers from the local land title companies, attorneys, or other genealogists) will be working there, also.

Use as little counter space as possible. If you have a coat, hat, umbrella, or briefcase, find a spot off the counter to place them; counter space is generally limited.

Finally ... It's in Hand

After you find the Deed book you seek, turn to the page you noted from the index. The document you want should start on the indexed page. Follow it through to the end, which may be several pages later. In Chapter 3 you learned about transcribing and abstracting. Use those techniques to get the information, or if you wish to have a copy of the entire document, ask the clerk about photocopying costs and procedures. The fees vary but are usually nominal.

Success at Last!

You are now looking at the deed. What to do? First, you need to read through the document. Until you become familiar with old handwriting and terminology, this will be a slow process. It's through such readings, however, that experience develops. You're seeking certain bits of information from the deed:

- Name of the parties (grantors and grantees)

- Residences of the parties

- Occupations of the parties

- Consideration "paid"

- A description of the property being sold

- Names of adjoining property owners mentioned

- Any special wording that may help in identifying the parties or the property

- Signatures or marks exactly as they appeared

- Names of witnesses exactly as they appeared

- Date and place the deed was acknowledged in court (or sometimes before a Notary Public if they moved to another area)

- Who acknowledged the deed

- Date of the deed, and date the deed was recorded

 TREE TIPS

Always carry a magnifying glass with you to the courthouse. The glass can help tremendously with writing that's difficult to read.

Copy names exactly as they appear (Chas., Rebeckah, and so on). Residences are important; they may even include mention of a former or later location: "John Gott, formerly of Lebanon, Connecticut" or "Richard Smith, now of Marion County, Indiana." Occupations are not only of personal interest, but can also be used to segregate the records of two people with the same name.

The "consideration" (that is, what was given by the buyer to the seller to obtain the property), may be monetary. It may also be something else of value, such as "love and affection" (plus a token amount of 5 shillings or $1), or five horses, or anything else the parties agreed upon.

The property description in the document will help you to locate the land. Additionally, it may provide clues through proximity. If someone of the same surname lived on the adjoining farm (mentioned in the property description within the deed), note it. Even if they didn't have the same surname, neighbors may have been related, or they may have been former neighbors who moved with the family from another area and thus can provide clues to the family's original location.

I Saw It Myself

All deeds had to be acknowledged or proved, either personally by the seller or by the witnesses who testified that they saw the seller sign the deed. If the seller moved and personally went into court to acknowledge the deed in another county or state, you have clues to a possible new residence.

LINEAGE LESSONS

The signatures in the transcribed deed books aren't the original signatures of your ancestors. The clerk copied them and often tried to duplicate the appearance. If the seller signed with an *X* (or other mark), the clerk tried to duplicate that, too. This may be important. Two men with the same name who left records in the same area can be distinguished by their signatures or marks. One may be able to sign his name, while the other always signed with an *X*.

The date of the deed is important, as well as the date it was recorded. Often a deed isn't recorded until years later. The delay can have special significance. Perhaps the father died and his widow received her *dower third*, with the remainder to go to the children. She continued to live on the property undisturbed, and after her death 10 years later, the children sold the property. A document's recordation date many years after the date on the document should alert you to a possible change in the family status: the death of the mother, a parent's remarriage, or children coming of age. Watch, too, for special clauses in the deed that may help with relationships: "I give and convey … the land I inherited from my father Joseph Schneider …." Important relationship clauses such as this are best quoted in your abstract so there are no misinterpretations.

DEFINITION

The law in some jurisdictions usually allowed the widow a **dower third** (or some other portion, depending upon state, time period, and other factors) in the land. She received a dower third upon the death of her husband (he could not will it away from her), but if he sold the property during his life, she usually had to sign a release of her dower interest. If she was unable to travel to the county seat, court-appointed representatives visited and questioned her, asking her whether the property was sold with her consent. This dower third didn't apply in all jurisdictions. In Connecticut, for example, at various times the wife had a dower interest only in the land her husband owned at the time of his death. Therefore, if he sold the land while he was living, she had no interest and wasn't required to sign.

Moving On

Work with the land records for as long as your time allows. They are one of the most valuable resources you can use, often showing relationships and yielding clues. If you don't have time to finish abstracting, you'll at least have the information from the deed index. You may find that the Family History Library has microfilmed some of the records you need, and after you return home you can order some of the film from a FamilySearch Center near you.

Those Departed Relatives

After you've tracked the deeds of your family in the courthouse, you'll want to do some research with *estates*. Estates are the whole of one's possessions, especially the property and debts left by a person at the time of death. They are another valuable source of clues. Relationships are often specified in documents related to family members' estates. Normally, those you first encounter will involve estates of decedents (that is, people who died).

There are other types of estates. Perhaps a minor inherited some property and the court appointed a guardian to manage his or her estate until the *age of majority*, or you may find the estate of an incompetent who was in need of a guardian or conservator.

DEFINITION

To reach the **age of majority** is to become of legal age. This usually was 21 years of age for a male, and 18 or 16 for a female, but the age differed from state to state and in different time periods.

More of Those Indexes

Sometimes the clerks have created a consolidated index of the early estates, which includes a variety of estate documents. There also may be individual indexes for wills, administrations, bonds, and other records. Sometimes the only indexes available are those in each individual book. Will Book A has its own index, Will Book B has its own, and so on.

Died With or Without

Basically, you seek two kinds of decedents' estates: testate and intestate. People who die with a will in place are referred to as dying testate, those who die without a will die intestate. If the person died leaving *real* or *personal property* that needed to be settled, an estate proceeding was filed in the county of residence. Some people, however, didn't have sufficient property (determined by the state laws) to necessitate a court proceeding, so you may never find an actual record.

> **DEFINITION**
>
> **Real property** is immovable property: land and, generally, what is erected or affixed to the land. **Personal property** is generally money, slaves, or goods (items that are movable and tangible). Animals, furniture, and merchandise are personal property. Virginia was an exception to this rule. Starting with an act in 1705, for a number of years under certain conditions, Virginia law treated slaves as real property. It was the only colony to do so.

Taking Charge

If your ancestor left a will, he or she usually named within the will an *executor* to handle his or her affairs. This created a *probate* proceeding. If the person died without a will, the court appointed an *administrator* to handle the estate; the process of distributing that person's possessions was referred to as the "administration." There are variations of administration: the court may appoint an administrator *CTA* (*cum testo annexo*, with the will annexed) to handle a will if the executor named didn't want to serve, or if the executor died or moved to another state or couldn't provide bond. You'll likely learn such important refinements as you use the records.

DEFINITION

Probate is the process of legally establishing the validity of a will. An **executor** is someone designated by the person making the will to handle his or her estate, as set out by the will, after the death of the **testator** (the person who made the will). An **administrator** is appointed by the court to handle the estate of a deceased person who has not left a will.

The Probate Process

Probate, the action to prove and admit a will, is initiated (usually by a relative or creditor) after the death of the testator. Notice is given by the clerk of the court that the will is going to be heard on a particular day; anyone contesting it may appear. At the time of the hearing, the court requires proof, by testimony of the witnesses, that the will was signed by the deceased and was signed of his or her own free will. If the court approves the probate, the will is transcribed by the clerk into a Will Book in much the same manner that a deed is transcribed into a Deed Book. The will is assigned a book and page number, and is indexed under the name of the deceased. The original will remains in the courthouse records; it isn't returned to the family, unlike an original deed. That original will, and other loose documents that are to be created in the following months, in many states comprise a "probate packet," which hopefully remains in the courthouse. If so, the index should have a column for File Number so you can locate the packet.

TREE TIPS

It may be worthwhile to have the complete probate packet photocopied. Sometimes, items from the packet are lost, or the entire packet itself can be mislaid or misfiled. Get them while you can!

Always examine the original loose papers in the probate packet, if they exist. Here you'll find the original will (important to your search if the clerk's transcribed copy in the Will Book has an error). If your ancestor could write, the original packet may reward you with his or her original signature. The heirs may have signed receipts for their portions of the estate, providing additional signatures. If an heir was a married woman, the receipts often give the name of her husband, because he signed "in right of his wife" since by law he controlled the couple's assets. You will find other valuable documents included in the packet.

But He Didn't Leave a Will ...

If the person died intestate (that is, without a will), the first record was usually when a relative (or creditor) came into court and requested permission to administer the estate. A variety of records may be generated from such actions, but these aren't always indexed. To find them, you may have to do a page-by-page search of the record books. Search also for an administration packet, similar to the probate packets. The index should have file numbers for these records.

Checking the Estate Records

Keep in mind that you are seeking one of two types of records: a testate estate (with a will), or an intestate estate (without a will), often called an administration proceeding.

Each type generates a variety of additional records. You may find a petition to initiate the estate process, bonds for the executor or administrator in case they don't fulfill their duties, inventories of the decedent's estate, accounts listing all that was owed and due to the estate, petitions for the sale of the real estate of the deceased, estate sales when the property of the estate was sold at public auction, distribution of the estate, receipts of the heirs for their portion, and others. Each can provide wonderful leads, and a rare insight into the lives of your ancestors. By the tools listed, you may discover clues to the trades of your ancestors, get a glimpse into their education by the books they owned, and find other intriguing bits of information.

Letting the Published Indexes Assist

The classic in published estate indexes is Clayton Torrence's *Virginia Wills and Administrations 1632–1800* (Baltimore: Genealogical Publishing Company, 1965), published as an index to the early estates of Virginia. Others have created indexes to some other states: the wills of North Carolina, the estates of Ohio, and so on. Look for them in libraries.

I Do Take Thee As My Lawful ...

Find out which office holds the marriage records. Ask to see the marriage indexes. In most cases, they'll be open for inspection, although some states don't allow access to the actual records. There may be a separate index for grooms and brides, or there may be an index only to males.

After you locate the book and page reference in the index, check to see if the marriage books are available for use. When you find the record you seek, take down all the pertinent information, including the names of the bride and groom, residences, ages, occupations, the date of the license, the date of the marriage, who performed the marriage, witnesses, consent if either was a minor, and so on.

Nothing on the record is insignificant; the smallest detail can lead to further records or identification of the individuals. The witnesses may be related to the bride or groom. If either the bride or the groom was a minor, a parent's or guardian's consent will be valuable. Finally, the name of the person officiating may help to identify a church and lead to further sources.

TREE TIPS

If there's only a groom index, check to see if the marriages of the county were published. If so, the published book should provide an index to brides.

Marriages generate a variety of documents: marriage bonds, licenses, applications, consents for the marriage of a minor, and certificates or "returns" showing that the marriage took place and who officiated. Sometimes these are in separate books with separate indexes, or they may be combined into the same record. What's available depends on the time period, laws, and local customs.

In addition to the bound marriage books, in which the clerk entered the details of the marriage, you may find original, loose marriage bonds. In earlier times, a bond was given by the intended groom. If he changed his mind and didn't marry the intended bride, his securities on the bond had to pay the amount of the bond. As an option to taking out a bond, he may have had the *banns of marriage* announced.

DEFINITION

The **banns of marriage** was an announcement, usually in church, of an intended marriage. Normally, it was announced successively for three weeks. In some areas, this was the prevailing custom. A marriage bond necessitated paying a fee to the clerk of the court, so the alternate method of "publishing" the banns of marriage was an attractive substitute. The banns may be noted in church minutes, but they aren't recorded in the county record books.

The Tip of the Iceberg

On subsequent trips to a courthouse, you'll likely find records you didn't think to examine the first time. Look at a document with a view toward deciding if there could be other documents connected with the same action. If so, pursue those, too; they may have the answer to the relationship you've been pursuing.

There is no substitute for hands-on use of the records. In the office of deeds, there may be survey books, plat books, mortgages, oil leases, power of attorney books, tax books, and a variety of other related records.

LINEAGE LESSONS

The laws vary as to the beginning date when birth and death registrations were required by each state. Prior to state registration, the counties normally maintained their own birth and death registers. Check for their availability at the courthouse. The information may be minimal compared to present-day records, but if they exist, they may be enormously useful.

In the probate office you'll find court minutes, court-order books, inventory books, bond books, account books, settlement books, estate packets, and others. If they've been preserved, you may also find the packets of the original court documents.

In the civil records office there'll be papers involving small and large claims: citations, debts, attachments, levies, summons, divorces, notices required to be published in the newspaper, and depositions, just to name a few.

For a detailed discussion of courthouses, see Christine Rose's *Courthouse Research for Family Historians: A Guide to Genealogical Treasures*. Rose discusses every phase of courthouse research, with numerous examples and an extensive glossary.

Are Courthouse Records on the Internet?

Every genealogist who started their research on the Internet must eventually go beyond what's available from that source. This is especially true when working with courthouse records. Though there are isolated cases of digitized court records being available online, the vast majority aren't accessible in that form. The Internet, however, can help in related ways.

A useful site is the State and Local Government on the Net (SLGN) website at statelocalgov.net. Experiment with the options. These websites won't have the actual records, but they'll provide some helpful aids, such as historical background, maps, hours, and location.

Another website is the Massachusetts Registry of Deeds at mass-doc.com/land_registry_dir.htm. This one takes you to a page with links to maps of the state's towns/cities and counties, plus links to counties. From there you can access the county courthouse information.

The site at co.ulster.ny.us is for the Ulster County, New York, local government website. Substitute the name of another county in the URL, such as co.ontario.ny.us and off you go to that county, where you'll find maps and more.

These websites are just a few examples. There's an extraordinary number of sites available to help you when researching local government.

Will these websites serve as a substitute for utilizing the records within a courthouse? No. Some of the sites may have digitized a few of the records and some may post a few indexes, but these resources are miniscule compared to what's available in a courthouse. In no case are all the valuable courthouse records, or even a majority of them, available on the Internet.

LINEAGE LESSONS

One of the biggest fallacies of genealogy is that "everything is available online" or on microfilm or in books. Not so—there's much, much more. Researchers posting messages often lament that they have a "brick wall" problem when in reality they have many clues to pursue; the answers just aren't online, and they aren't all in published books or microfilm. Ultimately, it will be necessary either to visit the courthouse personally, write them a letter, or hire someone to go there for you.

Each trip you make to the courthouses will bring new memories and experiences. You may be offered coffee or you may be ignored by the clerks. Either way, remain courteous and friendly, and thank the clerks when you leave. The impression you leave will influence treatment of the next genealogist who arrives. You'll remember each experience: enjoying the strawberry festival in the courthouse parking lot, going down three subbasements and through three locked doors to work in the old deeds, or finding another visitor there working on the same family. You may stop and consider, in awe, that you're actually holding in your hands a document that was written during the Revolutionary War, or a 1720 will signed by your ancestor dividing his few possessions. He was holding the very same paper you now have in your hand.

I'll bet you, too, will become addicted to genealogy!

The Least You Need to Know

- The books in the courthouse are large and heavy; bring someone to help if you can't lift them.
- Have an idea of which records you would like to access, and the time period you're seeking.
- Start your search with the indexes; use the references there to find the record book.
- Carefully abstract the records you find. Be alert for any clues to relationships.
- Many records are available *only* in the courthouse itself.
- If you can't go personally, write a letter!

A Picnic in the Cemetery

In This Chapter

- Exploring cemeteries for research
- Finding the records
- Locating the cemeteries
- Recording the information

Cemeteries are quiet, peaceful places for contemplation and remembrance. They're also an excellent source of genealogical information. Plot placement, tombstone inscriptions, and records can fill in blanks, lead you in new directions, or add insight to your knowledge of your ancestors as people.

Why should you visit the cemetery if you already have a death date? Because you never know what you'll find. Here are two children who died in infancy who you never knew about. Grandma's tombstone has an inscription that moves you to tears. Grandpa's marker is engraved with a Masonic symbol that suggests other records to check.

The Chicken or the Egg?

Cemetery research has two goals: to find the cemetery and to find the records created by the ending of life. Whether you look first for the cemetery or for the records is like the chicken and egg problem. Without knowing the cemetery, you'll have trouble locating the records; if you haven't found the records, you may not know which cemetery or where in the cemetery the grave is located. Different circumstances require different approaches. You decide the best approach based on the information you have.

The Usual Preparation

On your trip to the area where your ancestors resided, allow time to search for burial records and for the cemeteries. You're already prepared with the background information you need on the surnames and any variants, the approximate death dates, and the names associated with your family. You may have the name of a cemetery from a death certificate, an obituary, or interviews with family members. Your library research may have turned up cemetery surveys that list your ancestors, or you may have clues from a county history.

Courthouse research on the trip may turn up deeds so you can locate your ancestors' residences. Where your ancestors lived in the area often influenced where they were buried. Travel was limited, and people were usually buried close to the home place. In many cases, they were buried *on* the home place.

After you locate your ancestors, look at your county and topographic maps to find possible cemeteries. Often you can find numerous burial grounds within small geographic areas.

Cemeteries on the Web

Incorporate online resources into your work to help you identify cemeteries or gather information about a particular cemetery. To learn about cemeteries in a geographic area, try searching by state, county, city, or town. Search usgenweb.com and rootsweb.com lists.

If you have the name of a cemetery, check for a web presence. Cemetery websites range from the minimal to the expansive. Some have a history not only of the cemetery, but also of the entire area. Others are lavishly illustrated with drawings and photographs. The site for Cypress Lawn Cemetery in Colma, California, cypresslawn.com, has brief biographies of the notables buried there, a veritable who's who of early San Francisco and northern California. Many sites include detailed maps of the cemetery, whereas others offer access to records. The website for Maple Grove Cemetery (located in Wichita, Kansas), http://maplegrovecemetery.org, answers genealogical inquiries free of charge.

There are veterans' cemeteries, state hospital cemeteries, poor farm cemeteries, and slave cemeteries. There have been burials in cemeteries on Indian reservations, military bases, and in World War II internment camps. All manner of cemeteries have Internet sites. A search for *pioneer cemeteries* quickly yields well over 69,000 leads. Using *ghost-town cemeteries* as the parameter for a Google search results in at least 2,400 sites.

The government helps pinpoint veterans' gravesites with its national grave locator at cem.va.gov. Its lists of national and state veterans' cemeteries include contact information. You will find background details about Arlington National Cemetery and its rituals and traditions, as well as biographies of many who rest there, at http://arlingtoncemetery.net. A particularly interesting website is epodunk.com. Enter a state and then the name of a community. The resulting profile of the community is replete with statistical information and numerous links, one of which is to cemeteries in the community or within a few miles. Even tiny communities, such as Essex, Illinois, with its 2003 population of 651, have their own page with the latitude and longitude. Select the cemetery link to find the Essex cemeteries (2) or the county cemeteries (53 for Kankakee County). Clicking on the topographic map link takes you to a USGS quadrangle map at topozone.com, where you can see the exact location of the Essex cemeteries.

If your ancestor was involved in politics, check The Political Graveyard (politicalgraveyard.com). This ongoing collection includes close to 217,000 federal and state officeholders and candidates, party officials, judges, and some mayors.

TREE TIPS

Large cemeteries may have ethnic sections established by tradition, covenants, or discrimination. They may have a "Potter's Field" where the poor or unknown are buried; often no records are kept on the individuals buried in this section.

Kinds of Cemeteries

Both the cemetery and its records are important to your research, but they may not be in the same place. There are several kinds of cemeteries: public, private, family, religious, and fraternal. Everything about their records varies: the information, the location, and the accessibility. Cemeteries change hands; a commercial venture goes bankrupt and a municipality takes over, or the county can't afford the upkeep and sells the cemetery to a business enterprise. Cemeteries sometimes stand abandoned when all descendants in the family are gone, or when the town population is down to

200 residents who no longer bury their dead in the town cemetery and have no wish to keep it up.

- **Public cemeteries** are owned and maintained with taxpayer money by a governmental jurisdiction (county or town). Their records may be at the courthouse or city hall, or in an office on the cemetery grounds. Their records, occasionally difficult to locate, are usually open to the public.

- **Private cemeteries,** or memorial parks, are for-profit businesses. Although their records are private, most are willing to help researchers.

- **Family cemeteries** range from a few gravestones in a corner of the pasture to larger cemeteries, where not only extended family members were buried, but also others from the community who had close ties to the family. Family cemetery records—if they exist—are usually difficult to find; they may have been deposited at the library or historical society, or they may have been handed down through the family to someone who left the area. In some cases, the deed to the property made a provision to preserve the family burial ground.

- **Church cemeteries** may adjoin the church or may be located some distance away. Church burial registers may take some digging to find. They may be at the church, but in many cases, they moved with a minister or are archived at another location, such as a regional church archive or a university collection.

- **Fraternal organizations,** such as the International Order of Odd Fellows (I.O.O.F), also maintain cemeteries, sometimes adjoining another cemetery. A fraternal organization may also have a special section within a cemetery.

When the Cemetery Moves

Public works projects, such as dams and highways, sometimes make it necessary to move cemeteries, or the government may decide to consolidate several military cemeteries. In these cases, movers attempt to identify all the burials and move the remains to another location, but the original relationship of the graves to each other may be lost in the move. Occasionally, developers are unaware—or disregard—a small cemetery on a property, and obliterate all records, both on paper and in stone.

Burial space in the churchyard may become full, forcing the church to start a new cemetery. You may need to search both. Sometimes churches remodel and build over an old, unused cemetery. Churches merge or split; this, too, affects the cemeteries and their records. Sometimes the cemetery doesn't move, but its name changes.

Sexton's Records

Both public and private cemeteries have *sexton's* records. These are the records you want to see. Some cemeteries have a small sign at the entrance with instructions on how to reach the sexton. For others, call the listing in the telephone book. In very small towns or rural areas, there may be no telephone listing. Ask local people for the name of the records custodian. The records of cemeteries no longer in use or of abandoned cemeteries are somewhat hard to find. Check the libraries and historical societies for leads.

DEFINITION

A **sexton** is a caretaker responsible for burials and cemetery maintenance.

In a small cemetery, the sexton may do everything: dig the graves, cut the grass, maintain the records, and sell the plots. The records may be in his home. He can be a valuable resource if he has lived in the community a long time.

Large cemeteries usually have an office where the records are housed and maintained, but this doesn't always hold true. At one large cemetery I visited, the sexton maintains his records in card files at his florist shop across the street from the cemetery gates.

Sexton's records vary, but may include burial registers, plats, plot records, and deed records for the plots. You may also find records of grave openings, notations that a body was exhumed and shipped elsewhere, or a mention that someone else is buried in the same grave. Don't expect to find complete records, especially for long-existing cemeteries; it's unusual to be so lucky.

Burial Registers

You'll find burial registers in chronological order by the burial date, and they may be indexed. If not, you'll have to search through the lists based on the estimated death date. The burial register may have only the name, burial date, and plot. Others are more extensive, listing age, birth place, marital status, death date and place, and cause of death.

TREE TIPS

Genealogical information in the burial register is dependent on the knowledge of the informant. Always confirm the dates and places with other sources.

Plats

The plat maps of the cemetery show grave locations and plot ownership. Active cemeteries keep these up-to-date to know which plots are for sale. Usually there are no dates, and if the plot owner's name is different from the surname you're researching, these may not help you. The plats can help you figure out where in the cemetery the graves of interest to you are located.

Plot Records

Usually these are card files that include the name of the plot owner, date of purchase, and names and dates of burials in the plot. If the plot is in perpetual care, which requires a yearly maintenance fee, you may be fortunate enough to find a present-day descendant in the records. You can usually assume everyone buried in the plot is related, but assumptions can sometimes set back your research. When my husband and I visited the cemetery where his grandparents are buried, we saw a marker for someone no one in the family had mentioned. Suspecting a scandal of some sort, we asked my mother-in-law. She replied that Grandma was a kindly soul, so when a stranger in town had dropped dead and no one knew his kin, she had him buried in the family plot.

As with other land transactions, the cemetery plot owner receives a deed. The sexton records it in the cemetery records, and in some areas, it's also recorded in a special county deed book.

TREE TIPS

The plot records may give you the married name of a daughter who purchased the plot. They may also list individuals buried in the plot who have no markers.

Do You Need to Visit?

Yes, do visit! The sexton's records don't describe the monuments, nor do they tell you who is buried in proximity to your ancestors.

There's no substitute for strolling through the cemetery. Your ancestors walked this very ground. Grandma wept as she buried her third child during the smallpox epidemic. Grandpa chose to lay his parents to rest at the top of the hill overlooking the lake. Sons and daughters planted trees in loving memory—those same giant trees with roots now threatening to topple the monuments.

Locating the Cemetery

The libraries and historical societies in the counties of your ancestors may offer finding aids for the cemeteries. These aids vary from a simple map showing the major cemeteries in the county to printed abstracts of one cemetery's records. In a tiny one-room historical society in Illinois, there are dozens of binders for the cemeteries in three contiguous counties. Each binder has a short history of the cemetery, a map, photographs, and an indexed list of the tombstone inscriptions found there. Similar records exist elsewhere.

> **TREE TIPS**
>
> Civic groups may know cemetery locations from the work on spring "clean-up" days; mowing, fixing fences, removing leaves and weeds, and other chores to tidy the grounds. Farmers and hunters may know of "lost" or forgotten cemeteries they have come upon while wandering the land and nearby forests.

Follow the Money

Who paid the bill for the funeral? Funeral homes and morticians can be your biggest allies in finding the cemeteries and their records. They keep records, and usually they work enthusiastically with you to find information.

Traditionally, families return to the same morticians for all funerals, so the funeral directors are often well acquainted with many family members and may be able to refer you to relatives still in the area. They also know all, or most, of the cemeteries in their county and many in adjoining counties.

> **LINEAGE LESSONS**
>
> Although most funeral homes work willingly with you to find information, they aren't compelled to do so. Funeral homes are private businesses, sometimes owned by large corporations. In our litigious society, they may strive to protect themselves and their clients by treating all information as confidential. When seeking records, remain courteous and understand their position.

In modern times, the funeral director collects the information for the death certificate and the obituary. These records can give you birth dates and places, siblings, and children, as well as occupations and other personal information. Remember, the information was given by someone other than the deceased, so it may not be accurate.

The stress of the occasion may befuddle bereft survivors. The informant may be unrelated to the decedent, perhaps an in-law or a distant relative unfamiliar with the deceased's biographical details. Even if the death took place before death certificates were required, the funeral records are worth pursuing. For listings of morticians and funeral directors, consult *Funeral Home & Cemetery Directory* in book form or on a subscription basis at nomispublications.com/publications.aspx or the *National Directory of Morticians* which has free searches available at funeral-dir.com/directory/search.aspx. If a funeral home is no longer in business, contact other funeral homes in the area. They often know where the records are and may even have the records themselves. Conscious of the value of the records, many funeral directors make it a point to preserve old records whenever possible.

TREE TIPS

When funeral homes consolidate or become custodians of defunct morticians' records, they may archive old records in an off-site facility, so they may need to recoup their expenses by charging you for a search whether or not you find anything.

Procession to the Cemetery

The absolute best time to visit a cemetery is on Memorial Day, formerly known as Decoration Day. Traditionally, this is the day when all the family gathers to spruce up the burial grounds, plant shrubs, and put out fresh flowers and flags. Cousins play hide and seek among the monuments while the old folks reminisce about days gone by. I remember these gatherings well from my childhood, with my father trimming the grass around his grandparents' and baby sister's graves and my mother tending to her great-grandmother's plot.

All family cemeteries are different, from the imposing monument in the cemetery in the following figure of the Hon. Joseph Bush family in *History of Chenango and Madison Counties, New York* (Syracuse: D. Mason & Co., 1880) to a cemetery full of mostly field-stone markers surrounded by a barbed-wire fence. No matter the kind of cemetery or whether you go on Memorial Day or some other day, here are a few steps to success:

- Dress appropriately; you'll be out in nature.

- Take a pencil, paper, and digital camera.

- Plan your route to take in several cemeteries.

This sketch depicts the Bush family cemetery.

PEDIGREE PITFALLS

When you're actually ready to visit the cemetery, it is prudent to take someone with you. Cemeteries are often in isolated spots; use some caution. Your companion can help you hunt for names on the markers if you don't have a map of the cemetery.

No Fashion Statements

The well-dressed cemetery researcher wears long pants, long sleeves, and old shoes or, sensibly, boots. Cemeteries are well known for their abundant flora and fauna, especially chiggers, gnats, ticks, snakes, and small rodents. The grasses may be high and full of burrs, and the ground may be uneven and full of small depressions. Don't let these things deter you, but do be aware of the hazards.

A Picture Is Worth a Thousand Words

Whether you're a skilled photographer or strictly amateur, take photographs of the cemetery. Try to get a few panoramic shots, then focus on the plots, and then on the markers. Write a complete description of the cemetery: name, location (be explicit so you can find it again), and overall condition. Sketch the plots that are meaningful to your research and note their locations: "Between the first two lanes to the east of the entrance, middle row, large double granite marker facing west with the name

Harcourt. Nearby are Waggoners and Ballards." If you have the plot description from the records, use it; it's usually something like Section A, Block 6, Lot 2. If you have a global positioning system (GPS), record the longitude and latitude of the cemetery.

Mark Your Maps

You have a limited amount of time; make the most of it. Plan an efficient route so you don't waste time zigzagging back and forth throughout the county. If the cemeteries are on private property, get permission before opening the gate and crossing the field.

Engraved in Stone

The highlight of cemetery visits is reading the tombstones. The variety is astounding. From huge monuments to simple wood plaques, all were placed in loving memory of individuals who had strengths and weaknesses, as do we. They're teeming with information about our ancestors, if we'll only read them.

The inscription may be only a name and range of dates or it may be akin to a family group sheet in marble, with information on the parents on the front of the marker and all the children and their birth dates on the back. Relationships are engraved on the stones: "Wife of John A. Davis," "Son," or "Beloved husband." One of the most important finds may be the inscription on a woman's tombstone that says "Daughter of" and the names her parents. This may be the only record you'll find of her maiden name.

The stone you find may be only a chiseled stone as in the following figure (from a gravestone in the small family cemetery at *Bellevette* in Nelson County, Virginia).

Perhaps you'll find some sentiment. There may be a reference to a Bible passage, or the entire inscription may be in another language. Look carefully at any symbols or emblems on the tombstone. They represent membership in fraternal, patriotic, civic, religious, and veterans groups with records regarding your ancestors. Learn to recognize the more common ones, and sketch unfamiliar ones to research.

A chiseled stone stands in a family graveyard.

PEDIGREE PITFALLS

Be aware that some monuments are placed in memory of someone who is buried elsewhere. Monuments might also have been placed many years after a death, when the family was finally able to afford the cost. You may find family members or interested associates replacing crumbled markers with new plaques. In all of these cases of stones erected long after the burial, there's more chance of error in the information.

Reading the Markers Can Be Difficult

Sometimes, the bottom part of the marker has sunk into the ground or wind and weather have eroded the words. Weeds and tall grasses may completely obscure the marker. Dirt cakes in the letters; lichen creeps over the symbols.

For cases like these, it's useful to have a few tools with you: a trowel, grass clippers, a stiff brush (*never* a wire brush), rags, and clear water. With these, you may be able to clear away enough debris to make out the information.

In the recent past, genealogists were encouraged to take rubbings of the tombstones, but that practice is now out of favor. Tombstones are now an endangered species that need protection from the elements and genealogists.

PEDIGREE PITFALLS

Never use harsh chemicals or abrasives on tombstones. Their damage is irreparable. Attempt only gentle cleaning.

Sometimes You Are Disappointed

Age, environment, neglect, and vandals all take their toll on cemeteries. Many tombstones were made of sandstone and they are crumbling to dust. Years of freezing and thawing cracks the stones, and pieces are missing. Weather erodes the inscriptions. Vandals may have tipped over the markers. In one cemetery, dozens of stones were in a pile at the base of a tree. I could have cried as I realized there was no way I was going to know whether my ancestor's stone was in that big pile.

Don't Leave Yet

Look carefully at the tombstones in close proximity to your ancestor's burial place. Relatives were often buried in clusters, and you might recognize some names.

The tombstones tell of great sorrow as you find a family burying babies year after year. Look at the dates on tombstones throughout the cemetery. You may find many families burying children on the same days. Suspect an epidemic such as diphtheria or influenza. In 1918, "Spanish flu" decimated whole families.

A clue to the cause of death of women is often revealed on the tombstones: "Mary Smith, died 6 October 1878, age 22." Next to her is "Sarah, infant dau. of J. A. and Mary Smith, 7 October 1878." Mary no doubt died in childbirth.

Spend a little time communing with your ancestors. They want to be remembered. You are their ticket to immortality.

The Least You Need to Know

- Search both the cemetery and the records to get a complete picture of your ancestors.
- The Internet is a rich resource for locating cemeteries.
- Information in the records and on the tombstones is from an informant and is subject to error.
- Take a companion to the cemetery with you.

More Than News in the Newspaper

In This Chapter

- The differences among newspapers
- The variety of notices they carry
- The uses of indexes and inventories
- The value of notices

According to a local newspaper in 1855, your ancestor left the hometown to join a wagon train to Oregon. Another wrote home in 1862 after a major battle of the Civil War and made the local news with the details he related in that letter. In 1920, an old-timer wrote his lengthy reminiscences of the history of the town he helped settle in 1885. Newspapers are crammed with fascinating tidbits. You'll be enthralled with the news and the quaint notices preserved on newspaper pages.

Dailies, Weeklies, and More

Not all newspapers are the same. They differed then and differ now in frequency of publication and in focus. Here are just a few of the kinds of newspapers you will find:

- Daily newspapers: published in large communities
- Weekly newspapers: small-town newspapers or competitors to the local daily
- Ethnic newspapers in native languages; also African American newspapers that may include items not found in other newspapers.
- Religious newspapers
- Legal newspapers containing legal notices and court calendars

Seldom will you access the original newspapers. Because of their size, the fragile newsprint, and the scarcity of copies of early issues, access to the originals is usually restricted. Libraries across the country have microfilmed many of their holdings; these are often available on interlibrary loan.

Newspapers in the Area

Many newspapers have undergone numerous ownership changes. Others are no longer published, but their back issues may be preserved. Try the latest edition of *Gale Directory of Publications and Broadcasting Media* (formerly *Ayer Directory of Publications*). It will lead you to newspapers still in existence and provide background on their predecessors. If the newspaper you seek has gone out of business, it won't be listed in *Gale*, but it may be in a former *Ayer* list. See an 1880 directory online at archive.org/details/ayerdirectorypu00presgoog.

Some state libraries with extensive newspaper collections have compiled special lists by county and by town. Look for those in your library's reference section.

TREE TIPS

A guide to current bibliographies, indexes, abstracts, and other sources of value for newspapers is in Ancestry's *The Source*.

Checking for an Index

A surprising number of nineteenth-century newspapers have been indexed in recent years. Usually not every name is indexed, but certain subjects and the principal names are. Primarily the items indexed are those involving births, marriages, and deaths. Determine whether there's an index for the newspapers in your area of search. Try entering the name of that newspaper in your Internet browser and see what comes up. For example, I entered at random *Reading Pennsylvania newspaper* to see if anything would come up for Reading, Pennsylvania. I quickly found the *Reading Eagle*. A variety of articles were available online, including obituaries, some anniversaries with charming photos, and others. The newspapers on the web are often searchable for a variety of topics. Give it a try.

Even if an index hasn't been published for the newspaper, a local group may have created an unpublished manuscript or card index available at their library or genealogical society.

After you determine which newspapers were published in the area and whether they're indexed, you need to identify which repository has the newspapers and whether you can borrow the microfilmed issues on interlibrary loan. Ask your reference librarian for assistance.

Reading Every Word?

Nineteenth-century newspapers are difficult to read; often all the local items are combined into one long continuous column. It is hard to immediately locate the item you seek, unless you're fortunate enough to be working with a newspaper that headed its columns "Births," "Marriages," "Deaths," and so on. There's a shortcut to finding notices when indexes are lacking. Determine the name of the community in which your family lived. That leads you to the local columns. Community columns included diverse items all mingled together. Residents who took trips, visitors from out of town, who was ill, deaths, social events—all were listed one after the other without a break. The community notices can be particularly valuable when listing visiting relatives from out of town or noting the trips residents made to other localities to visit their kin.

Topics to Target

There's so much that can be helpful in genealogy, regardless of whether they're nineteenth- or twentieth-century newspapers. You may find these:

- Obituaries and death listings
- The family thank-you notice after a funeral
- Birth announcements
- Marriages and anniversaries
- Church news of members (particularly births, marriages, and deaths)
- Property sales
- Legal summons and citations
- Estate notices
- "Left my bed and board"
- Went west

- Advertisements

- Unclaimed mail at the post office

- Letters to the editor

- Visiting family and community events

TREE TIPS

When you're trying to get information from a county in which the courthouse burned, the newspaper can help fill some gaping holes.

He Died on the Fifteenth of June

Notices of death vary greatly, whether in current newspapers or those published 200 years ago. There may be only a brief mention in a column of deaths (usually called "Death Listings"). There may also be a full obituary including age and place of residence, a summary of the deceased's life, career, church affiliation, lodge membership, and much more. Always, without exception, seek these out. The names of the parents, where the deceased was born (providing the prior location of the family), when he or she was born, when and where he or she died, and the places the deceased lived all may appear in a full obituary. You may also learn about the deceased's marriages and the names of survivors, which can point to other places to search—brothers, sisters, children, and others connected to the family may be listed in the obituary. If the obituary lists the church where the funeral was held and the cemetery where your ancestor is buried, you may have even more potential sources. Occupation, professional career, war service, special skills (weaving, making quilts), town offices held, and other such gems may help round out the details of your ancestor's life.

Checking Several Papers

Don't limit yourself to one newspaper. Determine which newspapers were in the area and examine them all. In the weekly publications, check at least three to four weeks after the death date. Examine daily issues for at least a week later. Occasionally, the notice wasn't published for a significant time after the death date. On a routine basis, it's impractical to examine the newspapers for an extended period, but if the information is important to your search and you haven't been able to locate the notice, take the time. If the person died in a new location, check the old location, too. News often drifted back several weeks later to the original hometown, resulting in a notice published on the death of one of their own.

After you've located the notice, don't stop your research. Look at the next two or three issues. In its rush to publish, the newspaper may at first have scanty details and enlarge upon them the following day or two. I have found this to be true often enough to warrant a search for at least a few days beyond. If the newspaper is a weekly, check the issue for the following week. In the 6 December 1905 issue of the *St. Lawrence Republican* of Nicholville, New York, the obituary starts: "Hiram M. Rose whose death was mentioned *in last week's issue* ..." (italics added). The subsequent notice tells that he was born in Vermont and came to New York as a boy, and lists his various residences during his life as well as where his mother and father died. It includes his three marriages, the years, to whom, and the children born of each. If you had stopped with the first notice, you would have missed this.

If the deceased died on 4 August 1871, don't assume a newspaper of the same date would be too early. If you start searching with the following day, you may miss the notice. The newspapers were more flexible than present-day newspapers in their ability to add last-minute items, or that issue may have been late going to press.

Even a brief notice, such as the one that appeared in the *New York Herald* of Saturday 26 June 1852, can provide gems. The death of William E. Rose was reported, "after a short but severe illness, aged 27 years, 1 month, and 7 days. Relatives and friends of the family and members of Hook and Ladder Company No. 1 were invited to the funeral to take place from his late residence at 33 Forsythe Street." This provides an age at death from which you can calculate a birth, a residential address so you can check city directories for others of the surname at that address, possible land records (if he owned the residence), and even his occupation, which may lead to union or guild records in connection with his work.

Other Unexpected Rewards in Obituaries

Obituaries are immensely valuable in helping you to locate relatives who moved elsewhere. The list of survivors may include the brother who went west and the uncle who still lives in Boston. Others who traveled a distance (and may be related) may also be listed.

When you find the notice, try to get a photocopy (or a microprint if it's on microfilm). Note the date of the issue, the full name of the newspaper, and the page and column of the notice.

TREE TIPS

Many newspapers charge a fee for a listing; therefore, death listings aren't complete even in present times.

Urban Versus Small-Town Newspapers

Normally, the small-town newspaper was more expansive in its notices than the larger urban newspapers. Tight-knit communities wanted details. They knew the family intimately and could provide interesting bits of data. The sheer volume of people in populated areas mandated that only selected obituaries could be included—usually prominent individuals or long-time residents. This holds true today. It can be especially important to check for death listings in a city newspaper, because they publish so few full obituaries.

LINEAGE LESSONS

Some of the "best" obituaries are those published by religious societies. A fellow church-goer who knew the deceased would write what we as genealogists love—notices giving the whole family and obscure details we wouldn't find elsewhere. Church obituaries aren't always in the published religious newspaper; some are actually in the church's minutes. Determine the denomination, and do some investigating on the Internet for existing records.

The Family Thanks You

In some areas, it was—and still is—popular for the family to publish a card of thanks in the newspaper during the month following the funeral. Watch for these. A pair of examples follow.

These figures from early newspapers show two cards of thanks from the same family; one notice was published after the death of the father and the other after the death of the mother. Note the discrepancies in the lists. Some differences may be explained by marriages, but others are clear errors in one or the other list. Research will determine which names are correct. Such discrepancies are common and emphasize the necessity for gathering as many records as possible on the same individual.

The family may have published a notice to mark the anniversary of the death of their family member. Typically, these are brief "In Memoriam" types of notices. They're often signed, providing names of living relatives.

CARD OF THANKS

To those who have ministered unto our husband and father, J. M. Rose, and did all that loving hands could do, and to those who spoke of our aching hearts through the silent language of beautiful floral offerings, we turn with thankful hearts. May God of All comfort and bless you.

Mrs. J. M. Rose and Irl,
Mrs. Thirza Barr,
Mrs. Emma Wick
Mrs. Kannie Fields,
Otis Rose,
C. H. Rose,
R. L. Rose,
Ira Rose,
Harve Rose,
John Rose.

Here's the family's thank-you notice after the father's death.

CARD OF THANKS

We sincerely wish to express our thanks and appreciation to our many friends and neighbors for their kindness and sympathy to us in the sudden death of our mother.

Mrs. S. E. Wicker, Mrs. J. L. Gields, Mrs J .R. Barr, R. L Rose, Ira Rose, H A. Rose, O. R. Rose, C. H. Rose, John V. Rose, Irl Rose, Ora E. James.

Here's the family's thank-you that appeared after the mother died.

A Baby Was Born!

Some of the nineteenth-century newspapers published a special column of birth notices or mentioned them in community columns, although they weren't prevalent. Columns became more popular in the early- and mid-twentieth century. The columns were brief: "A daughter Mary was born to John and Martha Smith of Smith Twp." Such a notice may provide you with the first name of the mother if you didn't know it (and sometimes even her maiden name), the township, or other small bits of information you didn't have.

Wedding Vows

Marriage notices were—and still are—popular newspaper fodder. Perhaps news of the engagement was published, often with photos. Next, the couple may appear in a column of wedding licenses issued. This column usually lists the name of the intended bride and groom, their ages, and perhaps other significant details. After the wedding, there may be an article with a full description of the event and a photo.

The language and details or the editorial commentary in earlier newspapers were much more intimate than we see now. In a description of the wedding of one young woman, the newspaper reports, "It was intended that the father would give the bride away, but at the last moment he faltered, as it was more than he could do."

Silver and golden wedding announcements generate news. If you have the marriage date, add 25 or 50 years, determine where they may have been living, and check the newspaper. You may be rewarded with a photo and names and residences of close family members. There may even be a bonus: a wonderful description of the attire, the presents received, details of the original wedding, and relatives who came from afar to share in the celebration.

Christenings and More

If you can identify the family's religion, watch for church columns. The baptisms, confirmations, and other church news may provide you with another source of information on your family. Perhaps your grandfather was an elder, or your grandmother taught Sunday school.

Love Gone Awry

When a couple separated, the husband sometimes published a notice to absolve himself of legal responsibility for the wife's bills. We think of this as a more modern legal maneuver, but it actually was used very early. When Ezekiel Rose and his wife separated, he published a notice: "Whereas Mary Rose, the wife of me the subscriber, has left my bed and board, without any just cause, I therefore caution all persons trusting or in any manner dealing with her on my account, as I will not be answerable for any debt she may contract, or any dealing she may make, after this date." It was signed by Ezekiel Rose in Hampshire County, 15 March 1794, and published in the *Potowmac Guardian and Berkeley Advertiser* of Martinsburg, [West] Virginia, now preserved in the collection of The American Antiquarian Society. Without this notice, you might assume he was a widower when reading the will he made in 1818, omitting any provision for a wife, though Mary Rose actually survived him by 10 years.

In the late nineteenth century, in a moment of poetic inspiration, one husband submitted the following notice to *The Standard* of Jackson County, Ohio: "Mr. W. S. Williams of Illinois, announces that his wife, Ann Eliza, having left his bed and board without cause, he will not be responsible for any debts she may contract."

"Ann Eliza, Ann Eliza,

Once I loved but now despise her,

And So I no longer prize her,

I will go and advertise her,

For although I'm not a miser,

I won't pay for what she buys her."

LINEAGE LESSONS

In some jurisdictions in various time periods a published notice that the wife "left bed and board" was part of a divorce or separation proceeding.

Sale of Property

Property sale notices can be charming … and explicit. The executor of an estate is perhaps advertising the deceased's property, or the sheriff is selling a tract at public auction because of debt or taxes due.

In the following figure, the land commonly known as T. Rose's Old Place is advertised for sale in the *Maryland Herald* of Hagerstown, Maryland, on 3 March 1819. The description includes details that are just about impossible to find in other sources.

In a different matter, a lengthy advertisement was published in the 17 September 1817, issue of the *Adams Centinel* in Adams County, Pennsylvania. Being sold was the following:

> *Valuable Grist or Paper Mill Seat or any other kind of Water Works with 18 feet of head and fall, situated on Conowago Creek, in Franklin & Menallen townships, Adams county, three quarters mile from John Arendts Tavern, on the Road leading from Pine Grove to Gettysburg, with a LOT of 12 acres of land whereof 7 are excellent timothy meadow clear—the remainder is well covered with Timber. The improvements are a new two-story log dwellinghouse, with a back shed to it … for terms of sale, apply to the Subscriber living on the Premises.*

It was signed by John Mackley. The precise description in the advertisement enabled the family to locate the piece of land.

> THE subscriber offers for sale that well known Public Stand, commonly called T. Rose's Old Place, situated on the Potomac. The Cumberland Turnpike runs through the land nearly a mile, and passes immediately by the buildings. This is one of the best scites for a Tavern between Cumberland and Hagers-town, being forty one miles from the former and twenty-two from the latter place, and four from Hancock-town. There are about 40 acres of Potomac bottom, 12 acres of which are excellent meadow, and about 60 acres of upland, part cleared, the remainder in wood-land. The improvements are but indifferent. There is a tolerable good spring on the premises, and a run passing through, convenient to the improvements. It is presumed any persons wishing to purchase such property, will come and view it and judge for themselves. It will be divided to suit purchasers if required. The terms of sale will be, one half in hand; the other half in three annual payments.

Here appears an advertisement of land for sale.

Legal Notices: The Fine Print

Among newspaper items are the legal notices, those items usually in small print. They include items directed by law to be published to notify people who may have been interested in the action. Those items are often omitted among modern indexes, which is unfortunate. They can provide valuable leads in your quest.

When a defendant in a suit or heirs of a decedent can't be located, the law normally grants permission to publish the summons or citation to give notice to the parties involved. It may be published in more than one newspaper as directed by the court. Stop to examine these. You'll readily see their value.

TREE TIPS

In many areas, paper was scarce during the war years. The local newspaper may have suspended publication for the duration.

Other Miscellaneous Notices

During the gold rush to the west and other surges of expansion, the newspapers were packed with bulletins such as "John Smith, George Martin, and Gregory Morton left last Tuesday to join the train at Huntsville traveling west." Or, "Josiah Martin finished outfitting his team and wagon and left yesterday." The Civil War also generated many items about hometown boys who left for service, or news of when they wrote home.

Those Charming Advertisements

Was your ancestor a tailor? A pharmacist? Did he or she own a stable? Look for advertisements. They're delightful. The doctor extols the cures reported from the latest herbal wonder; the tailor confidently announces there's no workmanship that matches his own. Always make photocopies when you find advertisements placed by your ancestors. They add interest when you assemble the story of their lives, and the copies will add eye appeal when you illustrate your written account. (Remember, though, to consider copyright. Note the year of publication and determine if the copyright has run out by going to copyright.gov/circs/circ15a.pdf and reading about the duration of copyright. If it hasn't, get permission to include the image in your published document.)

A Letter Is Waiting For ...

Letters sent by anxious relatives, or others, often went unclaimed at the post office. The recipient was either unaware the letter had been sent, or had moved away. The newspaper periodically published lists of unclaimed mail; it may be the only proof your ancestor was supposed to be in the area. In other instances, a worried relative may have written a letter to the editor of the newspaper and inquired about "my brother who I have not heard from in over five years … have your readers heard of him … please have him write to …."

Ethnic and Religious Newspapers

If you can't find notices in spite of an exhaustive search of the English language newspapers, the information you seek may be in an ethnic newspaper. Was the family German and living in a large city? The item may be there. "But I don't read German," you may say. Doesn't matter. Watch for the name; you'll recognize it. If you find a notice, copy it and seek the assistance of a professor or student of the language at the local college or university. You can also seek assistance from an ethnic genealogical society to find others who can translate the notice for you. Some websites that offer translations can help, though personal experience with them has shown that these aren't always accurate. One you can try is babelfish.yahoo.com; another is translate.1888usa.com. They will at least give you a sense of what the article says, but supplement them with a true translation by someone knowledgeable.

Can the Internet Help?

You have a good chance of being lucky online, at least when researching some of your ancestors. Many organizations are digitizing newspapers and posting images. Some are searchable on every word. When the complete images are unavailable, there may at least be indexes, often prepared by volunteers. Some of the websites offering newspaper notices are subscription sites such as ancestry.com. Increasingly, libraries and genealogical societies are making large databases of newspapers available to their patrons and members. Go to the Godfrey Library website at godfrey.org for a number of newspapers, though there's a yearly fee for use. Others offering free archived newspaper access to members include the New England Historic Genealogical Society at americanancestors.org and genealogybank.com. There are many others. Start your newspaper search with Cyndi's List (cyndislist.com), but also experiment with your Internet browser, inserting the county, state, and the word *newspaper*. See what comes up.

To locate the website of many current newspapers, try usnpl.com. Once there, click on the state of your choice and peruse the links to the various websites for that state. Some of the current newspapers have archived past issues, though usually they're of a more recent time period.

Savor the Times

Use newspapers routinely during your search. They offer a rare opportunity to understand the times in which your family lived. Soak up the flavors of the area: the bake sales, local pageants, and sports-event winners. A strong feeling for the people and an understanding of the community atmosphere that influenced the lives of your family will be yours after reading those pages.

The Least You Need to Know

- There's much more than obituaries in newspapers to help in your search.
- Many newspapers have been indexed by general subjects.
- There are many published inventories to guide you to the locations of newspapers.
- It's hard to match newspapers for a flavor of the bygone era in which your ancestors lived.
- Some newspapers have been digitized and are available online.

Did Great-Grandpa Carry a Rifle?

In This Chapter

- Determining whether your ancestor served in the military
- Using Compiled Military Service Record files
- Reaping the benefits of pension files
- Discovering bounty land awards to soldiers
- Exploring the world of the World Wars' draft records

Knowing that an ancestor served his country instills in us a sense of pride. The diary revealing our fifth great-grandfather served with Custer causes us to rush to the history books for an account of those skirmishes with Indians. A newspaper account that another relation received a medal for his heroism at Gettysburg propels us to every website on that famous battle. The men who fought for our country helped shape our history. But how often can we really prove their service? Learn which companies they served in and what battles they fought. Descriptive records are there for the finding.

Didn't Know He Served?

My husband's great-great-grandfather was born about 1788 and didn't marry until 1819. Though he was about 25 and single when the War of 1812 erupted, no family tradition had indicated his service. Pursuing the possibility that he may have served, based only on age and the fact that he was unmarried, I contacted the National Archives. Completing the form they provided with my limited information on his age and places of residence, I requested a file. Imagine the thrill a few weeks later when a packet arrived with copies of papers signed by this ancestor, giving all the details of his service! His widow's documents in the same file gave information about their marriage, his death, and many interesting sidelights. This success so many years ago convinced me to always seek military files. You, too, will be more than a bit excited when you realize the extent of the records that exist if your ancestor was a veteran. Compiled service files, pensions, bounty land papers, hospital rolls, draft registrations, and many others await your discovery.

> **LINEAGE LESSONS**
>
> National Archives branches are located in Alaska, California, Colorado, Georgia, Illinois, Massachusetts, Missouri, New York, Ohio, Pennsylvania, Texas, and Washington. For specific locations go to archives.gov/locations.

The Treasure of the National Archives (NARA)

For military records before World War I, there's no greater repository than the National Archives and Records Administration (often referred to as NARA or the National Archives). With its principal facility in downtown Washington, D.C. (known as Archives I), and the facility in College Park, Maryland (known as Archives II), this repository offers many documents, microfilm, and maps that can assist the researcher. Many (but not all) of the National Archives records on earlier wars are available on microfilm. These and records of all Archives branches can be accessed at archives.gov.

Study the catalog titled *Military Service Records: A Select Catalog of National Archives Microfilm Publications* (called hereafter the *Catalog*), widely available in libraries and at all National Archives branches. Thousands of rolls of microfilm are listed. Some are indexes. Others are service records, unit histories, post returns, pensions, and more. The *Catalog* is available online at archives.gov/publications; click on "Online Publications," and then click on "Military Service Catalog." You'll use the *Catalog* often, so bookmark it. Better yet, invest a small amount to buy a copy from the National Archives.

You can order photocopies of original files using National Archives request forms. Use NATF-85, obtainable at Archives I, at any of its branches, or online at archives. gov/contact/inquire-form.html. There's an easier way, though: order and pay for the pension files and the compiled military files (as well as a few other listed files) online at the National Archives' website (eservices.archives.gov/orderonline). If the Archives doesn't locate the file you've requested, you're not charged.

The National Archives *Catalog* lists only the records on microfilm. Others, still in their original form (referred to as "textual records") are viewable in person in Washington, D.C., at Archives I, or you can order photocopies by mail.

> **TREE TIPS**
>
> Though the National Archives in Washington, D.C. has all the available microfilm, the Archives branches have only selected series. If the rolls you want are not available at an Archives branch, try other major repositories or a FamilySearch Center.

The Revolutionary War

Calculate the estimated age of your ancestor during the Revolutionary War. Don't consider 15, 16, and 17 too young; soldiers sometimes misrepresented their age to gain entrance into the military. Promises by the government to give land (in addition to pay) induced older men to try to enlist, too.

The Compiled Military Service Record

In 1800 and 1814, fires destroyed a large portion of the records of the American Army and Navy in the custody of the War Department. In a project begun in 1894, the War Department made abstracts from documents they purchased from a variety of sources. The War Department created individual packets for each soldier, and inserted the abstracted records into these packets. They also examined muster rolls, pay rolls, rank rolls, returns, hospital records, prison records, and others and then extracted the necessary information to bring together all the records relating to each soldier.

A typical Compiled Military Service Record, as it's known, gives the rank, military unit, date of entry into service, and whether discharged or separated by desertion, death, or other reasons. It may show age, birth place, and residence at the time of enlistment. There's no guarantee all of this information will be in an individual's packet, but normally they contain at least some of these bits. These compiled records are arranged by the war or the period of service, and thereunder by state (or some other designation), then by military unit, and finally alphabetically by the soldier's name. Those for the Revolutionary War not only have a microfilmed index, but also the complete original files are on microfilm. To access them you need only find a repository with both sets of microfilm or borrow the films from a FamilySearch Center.

Let's Try It Hands-On

Go to a National Archives branch or to a library that has the National Archives' microfilm. Look in the *Catalog;* it lists M860 with 58 rolls, and indicates the range of surnames on each roll.

As shown in the following figure, it's easy to ascertain which roll has the index for the surname you seek. Upon accessing the correct roll, find the soldier's name and copy all the information shown for him. With that information, you'll be able to find his file.

Roll	*Description*
1	A–Ange
2	Angi–Ballan
3	Ballar–Bearne
4	Bearnh–Biso
5	Biss–Box
6	Boy–Brown, Joh
7	Brown, Jon–Bur
8	Bus–Cartel
9	Carter–Chp

Here's an example of roll listings on microfilm publication M860 (by surname).

Searching for John Carter, go to Roll 9 of M860, as seen in the above figure. Find his name in the index on that roll. Copy all the information: regiment, company, and so on. Return again to the Catalog; it'll point you to a second microfilm publication, M881, the Compiled Service Records of Soldiers Who Served in the American Army During the Revolutionary War. Your file is on one of the 1,096 rolls. As seen in the following figure, the rolls in M881 are arranged by state, thereunder by regiment, and so on, and last by surname. Find the listed roll that matches the data shown by the index for the soldier. If John Carter served in Delaware in the 2nd Regiment of

New Castle County, Militia (determined from the index), his record would be on Roll 380 of M881. Now you can find the Compiled Military Service Record file. It may consist of only one card or a number of cards.

```
Delaware:
    380      1st Battalion (New Castle County)
             2d Battalion, Militia
             2d Regiment, Militia
             2d Regiment (New Castle County), Militia
               A–G
    381        H–W
             7th Battalion, Militia
             Hall's Regiment
    382        A–Br
    383        Bu–C
```

Here's the roll listings of microfilm publication M881, which contains the actual files.

A few of the rolls in microfilm publication M881 for the state of Delaware are shown here to illustrate how the rolls are listed. Match the information you found on your ancestor's indexed entry to determine the roll you need.

Now His Pension File

Don't be too disappointed if the Compiled Military Service Record is brief. At least you can now document that your ancestor served in the military. There are a multitude of additional records; one of the richest for genealogical information and interest are the soldiers' pension files often containing facts we value—birth date or age and birth place, marriages, and other similarly helpful details. The pension file indexes reveal whether your ancestor or his heirs ever applied for a pension.

The Act of 1818, based on financial need, is the first major pension act for which the application papers are preserved. It was quickly followed by an act in 1820 that tightened the requirements for pensions. As the government became more lenient, restrictions became more relaxed. In 1832, a general act awarded pensions based solely on six months or more of service. In 1836, widows received benefits. (Up to that time, only the widows of officers were eligible.) For more on the various pension acts passed by our government, see Christine Rose's *Military Pension Acts 1776–1858*.

The complete pension file for each Revolutionary War soldier is available on microfilm. Microfilm publication M804 of the National Archives consists of 2,670 rolls of alphabetically arranged records titled *Revolutionary War Pension and Bounty Land Warrant Application Files, 1800—1906*. (The dates shown are correct; applications were made by heirs as late as 1906.)

Microfilm publication M804 reproduces every paper in each pension file. Each soldier's file contains two groups: the "selected papers" (papers the National Archives considered the most important and which they previously used to fill mail orders), and the remainder of the file, marked as "nonselected." Examine every paper in your ancestor's file. Nonselected papers often include additional *affidavits* and forms; they may include letters from descendants around the 1920s and 1930s, when many (who wished to apply to a lineage society) sought information on their ancestors. These letters can lead to descendants. Though the National Archives used to provide only the "selected" papers for their minimum file, they now offer the complete file for a flat fee. Go to archives.gov for current pricing. If you have access to the microfilm, you can copy the papers and read them at your leisure.

DEFINITION

An **affidavit** is a written declaration made under oath before a notary public or another authorized official.

In another microfilm publication, M805, the Archives again reproduced the Revolutionary pension files, but in this one only the "selected" portion of the files were included in the microfilm. This shorter series was purchased by many libraries that couldn't afford the more extensive M804 series or didn't have space to store those voluminous rolls. If you find your ancestor on M805, make it a point to reexamine the file in its entirety when you can access M804.

Digital versions of the pension files for the Revolutionary War on M804 are now available at two subscription websites: fold3.com and ancestry.com. The site at fold3.com includes an index for almost all names (not just the veteran's name).

Now in Print

Another resource for your search of the Revolutionary War pension files is Virgil D. White's *Genealogical Abstracts of Revolutionary War Pension Files*. The abstracts are in three volumes, with an every-name index in a fourth volume. These abstracts don't include every paper in the files and aren't a substitute for examining the complete file. Nonetheless, they're invaluable in helping to determine whether your ancestor served.

TREE TIPS

The publication by Virgil D. White is particularly useful because these published abstracts are indexed in their entirety. Your ancestor may have made an affidavit in the application file of another soldier. You wouldn't find that affidavit without this index.

Why Check Further?

Why bother to proceed to the pension file if you already know from the Compiled Military Service Record file that your ancestor served? In short, because you'll learn a lot more from the pension file. You'll experience a connection to your ancestors as you read the words they spoke, detailing the battles in which they were engaged and the resultant disabilities and hardships.

You'll read about some sad situations, such as that of the widow Rebecca Rose, who at the age of 91 was found in the poorhouse, blind, "nearly naked, entirely helpless," defrauded by two unscrupulous men who filed her pension for her and gave her little of the funds. A man of conscience in the county came to her aid, demanding federal government assistance for this aged widow.

James Rose, the husband of Rebecca, a Virginian, had his share of difficulties, too. He tells that he was at Mill Creek when "the picket guard came in great haste, scared nearly to death," bringing a report that thousands of British were coming, just on the other side of Mill Creek Island. "Col. Mazzard having no horses at that time to manage the cannon, commanded the army to hasten to Mill Creek, and draw with them three of the cannons. This soldier [Rose] was one of the number that managed the cannon in the stead of horses, and produced a rupture in his body of which he never has and never will recover by his great exertions in drawing ...." Later, he was discharged and returned to King George County. He left his discharge at the home of a friend and, on a borrowed horse, went to see his relatives in King William County. On the way he was taken up as a deserter "by a company of drunkards" and retained in custody three days before he could get his discharge, "which he procured by giving a man a regimental coat to take the horse back and bring the discharge."

James' troubles weren't over; all are recited in the voluminous file. Fortunately for the unlucky James, his service was substantiated, and later a special Act was passed by Congress on his behalf. Similar stories abound. The files are fascinating and give you a rare opportunity to know your ancestors.

Here's a bonus: your ancestor may be one of the soldiers who either tore out the pages from the family Bible and offered them in support of the statements, or had the Bible entries extracted and notarized. Those papers may be your only proof of dates and relationships in the family.

The Compiled Military Service Record file and the pension file aren't the only records available for Revolutionary War service. There are many others, but these will get you started.

The Lure of Bounty Land

Bounty land was awarded by the federal government either as an inducement to serve or as an inducement to remain longer in service. Later it was a reward for service. Government legislation authorized bounty land (that is, a free right to government land) in 1776, and in a series of subsequent acts the land and requirements were defined.

DEFINITION

Bounty land is federal land awarded (under certain conditions) by the colonial and later federal government to veterans for their military service. Soldiers were promised this land either to entice them to join the service, to extend their enlistment, or as a reward for having served.

From 1788 the soldier could sell bounty land warrants from the Revolutionary War. Because it wasn't until 1796 when the United States Military District in Ohio opened where the soldier could exchange the federal warrant for land on which he could settle, most soldiers were persuaded to sell their warrants (usually to speculators); few warrants were actually located on the land. This defeated the government's hope that those veterans would settle on the frontier and help protect it.

LINEAGE LESSONS

The National Archives' microfilm series M829 contains warrants issued under the act of 1788 and other early acts. Indexes on the first roll lead researchers to the appropriate roll, which contains information pertaining to the deposition of the warrant issued for service.

During the War of 1812, the federal government imposed restrictions on the disposition of the warrant. The soldier could not sell or assign the warrant, but his or her family could inherit it when he died. This wasn't changed until 1852, when new federal legislation allowed individuals to sell unused warrants. Warrants issued for the War of 1812 were restricted to military tracts in Arkansas, Illinois, and Missouri until an act passed in 1842 allowed use on other available federal land.

LINEAGE LESSONS

The Virginia Military District, available only to Virginians who served in the Continental Army, opened in Ohio in 1794.

Many early bounty land application files were destroyed in the War Department fires in 1800 and 1814. Any papers remaining were combined with the Revolutionary War pensions and saved on microfilm in M804 (described previously). Bounty land warrant files for the War of 1812, under acts of 1811–1816, were microfilmed on M848 on 14 rolls and titled "War of 1812 Military Bounty Land Warrants, 1815–1858." The dates refer to the period within which the applications were made. The indexes are on the first roll of this series.

The bulk of the bounty land application files aren't on microfilm yet. Known as the "unindexed bounty land application files," they're based on acts of 1847–1855; no full name index is yet available. This mass of files resulted from more lenient acts that reduced the required length of service for eligibility and instituted other changes. The files are arranged alphabetically by the soldier's name in thousands of legal-sized manila files. These files are packed with items of genealogical importance: marriages, Bible records, death records, and many more. An index is being prepared by volunteers, though it's a long-term project. In the meantime, the completed part of the index is available only on a computer located in Archives I in Washington, D.C.

TREE TIPS

Though the early bounty land application files for the Revolutionary War were destroyed by fire, information from warrant registers and other related items are scattered throughout a number of repositories and states.

If you ascertain that you had an ancestor who served in the War of 1812, Indian Wars before 1855, or the Mexican War (or any service 1790–1855), fill out the National Archives form NATF-85, and mark the box on that form for bounty land file. The National Archives will search for a record when it receives your request. Keep in mind that the last act granting bounty land was passed in 1855; applicants who filed later were doing so for service prior to the passage of this final 1855 act, before the Civil War. Researchers should consult Christine Rose's *Military Bounty Land 1776–1855* for a full discussion of bounty land including application files, warrants, and laws.

The War Between the States

The Civil War tore families apart. Brother fought against brother, Families were divided, and many familial wounds never healed. Feelings run deep to this day. Be aware of this and be sensitive as you query your family about their Civil War connections. In one file, a Tennessean, faced with the decision of whether to stay with the

Union or join the Confederacy, moved to Arkansas to join a Union force there. His wife, whose family held staunchly to their Southern sympathies, refused to accompany him. Under these circumstances, the soldier later took a new wife in his chosen state—in this case without benefit of a divorce from the former wife who refused to live with him. The files are full of sad tales. Don't be surprised at what you may find. Remember the times in which these families lived.

It not unusual to find records of the same soldier serving on both sides. The 19-year-old who went into town to run some errands for his family was spotted by recruiters and forced to enter the service. At his first opportunity, he deserted and joined the opposing side, where his sympathies lay. If your ancestor was from a border state such as Tennessee, Kentucky, Arkansas, or Missouri, you're especially apt to find dual records for him.

LINEAGE LESSONS

Don't assume your ancestor's record stating "deserted" means he was a "coward." There were many reasons for desertion: loyalty to the other side, going home for a few days because of sickness in the family and then not returning, and more. He may even have been listed improperly as a deserter when he fell into the hands of the opposing army.

He Wore Gray

Start your search for the Confederate soldier with the National Archives' microfilm publication M253, *Consolidated Index to Compiled Service Records of the Confederate Soldiers*. Consisting of 535 rolls, this is arranged alphabetically by surname. When you locate the surname in the index, take down all the listed information.

An easier way to examine M253 is to go to ancestry.com where scanned images of this index are available.

Next, return to the *Catalog*. Look for the microfilm publication number for the Confederate Compiled Military Service Records of the state from which your ancestor served. The state rolls are arranged by regiment, then unit, and so on, and finally by surname. It should be fairly easy to determine the correct roll. You can then view on the microfilm the entire file for your Confederate ancestor.

Confederate Pension Records

In the years after the war, a number of states granted pensions to veterans of the Confederacy. These weren't federal pensions and aren't in the National Archives. These pensions were granted by the states to the Confederate soldiers who resided in their states at the time of their pension application. A soldier who served from Louisiana, but who later lived in Texas, filed for his pension in Texas. Some states have published indexes to their Confederate pensions. Typically, the indexes are available at the state archives or state historical society of the states granting the pension. Those same repositories usually hold the originals as well, so you can order a photocopy.

TREE TIPS

Some states have copied their Confederate pension application files onto microfilm. The state repository can advise you. Be sure to check their indexes for the widow's name, too; she may be the applicant. There may be two files: one for the soldier and a later one for the widow.

He Wore Blue

There's no consolidated microfilmed index to the Compiled Military Service Record files for those who served with the Union. Indexes exist on a state-by-state basis only. In the *Catalog*, look for the state from which your ancestor served; next look for your ancestor's name in that state's list for the proper film number. Copy all information shown for your ancestor—unit, regiment, or anything else, exactly as it appears. Doing so will lead you to either his original record (still in textual form) or to a series of microfilm that contains the file. If you don't find your ancestor but feel sure he served, then check the surrounding states' records.

After you've located your ancestor and copied the information from the index, determine through the *Catalog* whether the actual files are available on microfilm. Some of the Union service record files are on microfilm, but others (depending upon the state from which the man served) are available only in their original form. You'll have to view any original files personally, either by ordering them online or by mail (described previously). Follow the National Archives' instructions for form submission. If you order the file online, you can arrange for payment at the same time. For the fee schedule, go to archives.gov/research/order/fees.html.

If He Got a Pension

The chances are good that if he lived long enough, your Union Civil War ancestor applied for a federal pension. Five acts between 1862 and 1907 provided pensions based on Union service. The earlier acts provided for those who were disabled or killed. A widow, children, or a parent who could prove that an unmarried son contributed significantly to his or her support may have applied soon after the death of a soldier in service.

To locate a Union pension file, consult the National Archives' microfilm publication T288, *General Index to Pension Files, 1881–1934.* This index includes service for the Civil War and, in some instances, earlier war service by a Civil War veteran. Other entries in the same film series relate to service in the Spanish-American War, the Philippine Insurrection, the Boxer Rebellion, and those who enlisted in the regular Army, Navy, and Marines before World War I. The index is arranged by surname, and then by the state from which the soldier served. Be sure to take down all the information; you'll need this whether ordering the file by mail or viewing it in person at the National Archives. Images from this index are available at ancestry.com.

To order the Civil War pension file, use the same procedure as described for ordering Revolutionary War pension files. You can do it by mail using Form NATF-85, or online.

Some of the Civil War Union pension files are in the custody of the Veterans Administration (VA). If that's the case, the National Archives will respond to your request by sending you the appropriate VA office address to which you must write to request the information.

Though you can get a Civil War Pension file with only a few pages for a smaller fee, get the full file using the fee schedule already mentioned—otherwise, you are likely to miss important information. The Civil War pension files are routinely 40, 50, or 100 pages, or even more. The shorter version may save some money but you may miss crucial clues. Perhaps other researchers of the family will chip in for the cost and you can make copies for those who contributed.

Other Civil War Records

Two websites, fold3.com and ancestry.com, include images from the Civil War records. On fold3.com, you can find Civil War Maps, Civil War Widows Pensions, Confederate Amnesty Papers, and a variety of other records that are in various stages of scanning. Ancestry.com includes scans of T288 (the General Service Index from the Civil War through about 1900), U.S. Civil War POW Records, U.S. Colored Troops Military, and much more.

Twentieth-Century Conflicts

What if your research hasn't taken you back far enough to benefit from the wonderful records created from the Revolutionary to the Civil War? Don't despair. World Wars I and II, the Korean War, and the Vietnam War all generated records of interest to genealogists. As the memories of battles fade and participants die, the records become available, some even online. Go to cyndislist.com and click on "Military" on the "Categories" page to find many links.

TREE TIPS

If your male immigrant ancestor was born between 1873 and 1900 and was in the United States in 1917–1918, he had to register for the draft even though he wasn't required to serve. Draft cards can be an important source of information on your immigrant ancestor.

World War I Draft Registration

In 1917–1918, about 24 million men completed draft registration cards. This included aliens, who weren't subject to induction, but were required to register. This created a vast database of men born between 1873 and 1900.

These records are immensely helpful, providing all sorts of personal data. Birth date, birth place, citizenship, dependents, occupation, marital status, father's birth place—these things and more may appear depending upon which draft call your ancestor answered. To see what you can glean from these records, go to ancestry.com/save/charts/WWI.htm for blank draft forms of the following calls for registration:

- 5 June 1917: All men between ages 21 and 31

- 5 June 1918: All men who reached age 21 after 5 June 1917, and supplemental registration 24 August 1918 for those reaching age 21 after 5 June 1918

- 12 September 1918: All men ages 18 through 21 and 31 through 45

The original records are housed at the National Archives-Southeast Region, Georgia. Microfilms are available through the Family History Library, and ancestry.com has posted images. For an extended description of the World War I Draft with links to explanatory articles, go to http://search.ancestry.com/search/db.aspx?dbid=6482.

World War II Draft Records

This database contains an index and images of the draft cards from the Fourth Registration, the only registration currently available because of privacy laws. Referred to as the "old man's registration," it was conducted 27 April 1942 and required the registration of all men born on or between 28 April 1877 and 16 February 1897 (that is, between ages 45 and 64) who were not yet in the military. Images from their draft cards are included in the records. These records are available at ancestry.com. Other World War II items are listed at ancestry.com and at fold3.com.

Personnel Records

The National Archives in Washington, D.C., doesn't house personnel records of World War I and later. These are at the National Personnel Records Center in St. Louis, Missouri. If you're considered the next of kin of a deceased veteran (father, mother, spouse, sibling, or child), you can request a file on the special online form provided at archives.gov/veterans/eveutrecs. You can order other forms from archives.gov/contact/inquire-form.html. If you aren't the veteran or the next of kin, specify "Standard Form 180" to order records. Read the statements at the National Archives' website relating to privacy issues, and what you can order under the Freedom of Information Act. Experiment with the various links for other pertinent information.

A devastating fire in 1973 destroyed many records, but not all were lost. About 80 percent of the personnel records of those discharged from the U.S. Army on or between 1 November 1912, and 1 January 1960, have survived. About 75 percent of U.S. Coast Guard personnel discharged 25 September 1947, to 1 January 1964 (whose names fell alphabetically after James E. Hubbard), are extant. The U.S. Navy and U.S. Marine Corps records are intact. Since the fire, however, the National Archives has been collecting auxiliary records to help supplement the loss.

Searching for Details

Start your twentieth-century search for a relative's records by going to one of the following National Archives webpages. Each has several links. Explore.

- World War I: archives.gov/research/military/ww1/index.html

- World War II: archives.gov/research/military/ww2/index.html

- Korean War: archives.gov/research/military/korean-war/index.html

- Vietnam War: archives.gov/research/military/vietnam-war/index.html

Also check cyndislist.com for listings under military records. There are general sites listed, and others listed specifically by war.

Other Ways to Locate Evidence of Military Service

Your ancestor's gravesite may have a flag or marker indicating service, as may his tombstone. His obituary or death record may mention service. For the Civil War, the discharge may be recorded in the courthouse. The 1910 census records, which include a column for Civil War service, may list him; the 1890 special federal Civil War census records (discussed in Chapter 11) also may list him. In later wars, the separation papers may be recorded in lieu of the discharge. County histories are another source, often including rosters of those who served from the county. Lineage society records, many published, may also assist.

Virgil D. White, mentioned previously for his abstracts of Revolutionary War pension files, has published a series of indexes for pension files of other wars, and a variety of other military records. Look for these at your library. And don't overlook state records, including the Adjutant General's office, the state library, and special state projects (such as graves registration for soldiers or biographical sketches of all known soldiers of the state). Sources for military service are endless.

There's More?

This introduction to the military files of the National Archives acquaints you with some basic records created in the establishment and preservation of the United States. There are hundreds more files and records available: enlistment papers,

Quartermaster records, headstone applications, post returns, applications for military academies, and much more. The *Catalog* can give you an idea of the number of records available. Remember that the *Catalog* lists only the microfilm, however, which is a small part of the vast records created by military service. The suggested books will lead you to a multitude of additional sources.

There's a fascination in this country with the history of the wars and those who served. You'll find information surprisingly abundant once you have started your search. Don't skip these records "because I already know he served." Details—physical description, age, occupation—these and much more are yours for taking the time to understand these amazingly informative records. For the actual records of your ancestor's service, the National Archives is the best resource, whether the records are textural, filmed, or electronic. The web shines at its best when providing historical background and images. You'll find photographs of ships, planes, battlegrounds, uniforms, medals, national cemeteries, interactive maps, and servicemen; you'll read battle details and the memoirs of those who served.

The Least You Need to Know

- Always, without fail, seek military records for your ancestors.
- The National Archives has thousands of rolls of filmed military records and thousands of unfilmed files, manuscripts, and records books.
- Military records can be as brief as the period of service or comprehensive enough to give you three generations of ancestors.
- Such diverse items as physical description, marriages, deaths, Bible records, and others may be found in military records.

Making Sense of It All

You're going to collect a whole lot of records before you get through. You'll be cramming files into closets and under beds unless you learn to keep control of the paper mountains from the beginning. In this part, learn how to keep everything organized before it gets out of hand. You'll record your information with citations and effective numbering systems so it will make sense to you and to others.

You'll learn to make the records talk to you so your ancestors will come alive. They will no longer be just names in dusty records, but real people with personalities. Use the World Wide Web and many other resources to begin your understanding of the times in which they lived. You'll become aware of conflicting records and how to make judgments to resolve those conflicts.

Order Out of Chaos

In This Chapter

- Using binders effectively
- Establishing an expandable filing system
- Using the computer for maximum benefits

It won't be long before you discover one of the major problems of genealogy: the amount of paper generated by your research. If you work on several family lines, as most researchers do, you'll quickly become confused without proper organization. "Where are the notes from Aunt Mattie's interview?" you'll exclaim in despair.

The charts and family sheets that help record what you find are discussed in Chapter 4. Knowing how to use the charts and family sheets, however, isn't enough to keep you from being flattened under a mound of paper. You don't want to throw away anything important. "What do I do?" you lament.

If you're new to genealogy, establish a good filing system right away and use a computer to facilitate storing and searching. You have an advantage; you can prevent the massive paper problem before it engulfs you in a sea of paper. If you've been into genealogy for a while and you're already swimming in letters and documents, allocate some time to getting your files onto computer discs or flash drives. Not only will you be able to breathe again, but doing so also will help ensure your collection isn't later thrown out because of sheer size.

Filing the Charts and Sheets

When I first started in genealogy, I made no attempt to keep pedigree charts together. There were only a few, and there was no problem. But it wasn't long before I was hunting all the time. Where is the Thompson chart? The Burney chart? A three-ring binder proved to be the answer to my organizational predicament. Store your pedigree charts (see Chapter 4) in a binder numerically, starting with yourself as number 1 on chart 1. Assign each pedigree chart a number. As you leave one chart and start another, number the new chart, too.

After filing the pedigree charts, create a series of binders to file the family group sheets you've accumulated. Assign one binder per surname. If the sheets for a particular surname overflow their binder, continue into a second binder, still maintaining the filing system by family name. If you're tracing the family of John Jordan, who married Martha Adams, you would create a binder for the family group sheets of the Jordan family and another for the Adams family.

Filing Systems

Using binders for your pedigree charts and family sheets is a start, but you'll accumulate considerable additional paper. You need a simple system that you can expand easily as your search progresses. Create a set of file folders. Use either standard (8½ × 11) or legal-size (8½ × 14) folders. Legal-size folders may seem the most desirable because many document photocopies will likely be on legal-size paper, but legal-size filing cabinets are wider than standard-size, which can lead to significantly higher cost and more space needed if you eventually need two or three filing cabinets. It may be best to stick with standard-size folders and just fold the legal-size documents you need to file.

Now set up a system for filing the papers—all those letters, photocopies of census records, documents, biographies, and so on that you've been gathering systematically. Consider creating files by various categories:

- Family name
- State
- County
- Subject
- Correspondence

If you don't have enough papers in the file to create subfiles (shown in the following figure), then don't. Generate new files only as the need arises.

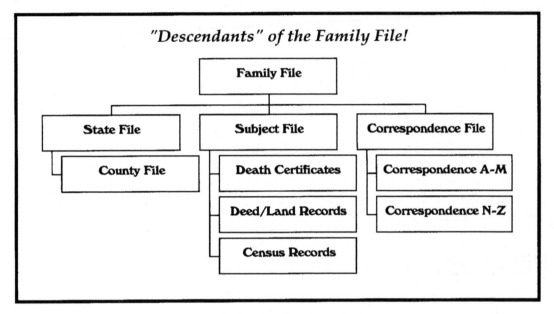

"Descendants" of the Family File!

Here is a sample flow chart of files.

When you start your search, set up family files for each of the surnames you're tracing. In each of these files, insert copies of letters you write regarding that surname and any documents you find involving that name: the deed, the marriage record, the Bible record, and so on. Anything that involves that family should go into that family's file.

They Called Tennessee Home ... and North Carolina

If you get deeply involved in one surname, you'll find that one file won't be sufficient for handling all the paper. It won't be sufficient to simply create a second family file for the same surname. In this case, you would have to search two family files for the record you want, and later three, and then four. Resist the temptation to start that second file; instead, establish some specialized subfiles, starting with state files.

Suppose you are working on the Martins. This family has become a focal point. The major part of your energy is devoted to them. As is typical of many early families, they lived in several states, defying your erroneous belief that your ancestors weren't mobile. Your file is overflowing with census records from Tennessee, deeds and tax lists from North Carolina, and a raft of documents from Virginia. Now is the time to set up state files. For each state for which you have perhaps 10 or more items for the Martin family, create a state file. Label it as "Martin—Virginia," "Martin—Tennessee," or "Martin—North Carolina." Also create a "Martin—Miscellaneous State File." Anything that doesn't fit into the first three can go into the miscellaneous folder. When you've accumulated sufficient records on Martin in another state, perhaps Pennsylvania, create another state file for that state.

When creating the state files, don't label them solely as "Virginia" or "Ohio." Later you may need to establish state files for records of another surname, so that label would be confusing. Always include the surname on the label, such as "Martin—Virginia." Be sure the surname is first, followed by the state, and file the folders alphabetically. To keep all your Martin folders filed properly together, begin the file name with the word "Martin," followed by a one- or two-word description of the file.

TREE TIPS

Consider using color-coded file labels or colored file folders. Assign a different color to each family name, each type of file, or each state. Perhaps all Pennsylvania files could be pink, all vital records files would be blue, and so on.

County Files

As the search progresses, even the state file may not be sufficient. If it gets too full, it will become unwieldy. You have a state file for "Martin—Ohio" but you find that your family lived in Pickaway County, Ohio, for a long time. They left so many records in Pickaway County that they take up half your "Martin—Ohio" file. Now is the time to create a subfile called "Martin—Ohio—Pickaway County." Move all the items in the "Martin—Ohio" file that relate to Pickaway County into this new file. Now when you want to examine the deed to the land in Pickaway County or verify that you have a copy of the estate papers filed in that county, you know immediately where to look.

Remember, when creating the new file for the Martin Family in Pickaway County, Ohio, it should be labeled as "Martin—Ohio—Pickaway County" to preserve an alphabetical filing system that keeps all the Martin files together. Always label your files with the broadest classification first, followed by the subject subdivisions, in order. If you're breaking down files by 1. Surname, 2. State, and 3. County, then the file label should be so marked.

Document Files

When you get several of a particular type of document on the surname, you can create a subfile for those specific documents, such as "Martin—Death Certificates." This time don't do it by geographic location; all of the Martin death certificates should go into this document file. On the family group sheet you maintain on the family, note that the information came from the death certificate. When you want to re-examine the certificate of John Martin, you'll know exactly where to find it. Similarly, if you have a number of newspaper items—obits, marriage notices, published legal notices, and so on—create a special "Martin—Newspaper Items" file.

TREE TIPS

It helps to file the certificates in your special death records file alphabetically by given name for easy retrieval. Not enough death certificates to have a file by itself? Name a file instead "Martin—Vital Records" and include birth, marriage, and death records.

Correspondence Files

At first, you may file correspondence relating to a particular surname in your basic family file, mentioned earlier in this chapter. But if you do considerable research on the family, you'll have correspondence mixed in with your other notes and papers. It is usually best to reserve the family file for research papers: the published biography, the manuscript written by your grandfather 40 years ago, and others. You need a way to handle filing your correspondence. Create a set of correspondence files. At first you'll need only one for the surname: "Martin—Correspondence." As the file gets larger, you can separate it into two files, "Martin—Correspondence A–M" for all your correspondents with surnames A through M who have written to you on the Martin family, and "Martin—Correspondence N–Z" for the rest. You can break it down even further as needed (A–C, D–G, and so on). If you have a great deal of correspondence with one individual, you may want another file devoted just to correspondence with that individual. Label it "Martin—Correspondence—Steve Stark."

Correspondence Cards

In addition to the correspondence files, you need an easy and effective index. One way to keep track of your correspondence, no matter which family it involves, is to keep a master correspondence card or log.

The card shown in the next figure can be either a 4 × 6 card, or an electronic log on the computer. Each entry should include the name and address of your correspondent, phone numbers (home and office), a fax number, and an email address. Also include the name of the spouse (if you know it). If a correspondent moves or changes telephone numbers and the phone number is listed under the spouse's name, you may need it to obtain the new number. Leave room on the card or in the log for comments, where you should include the specific branch of the family that the correspondent is tracing.

The very last line of each entry should be "SEE FILE." This is one of the most important items on the index card or log. Here you enter the name of the family file in which you filed their correspondence. If Mary Adams is tracing the Martin family and you filed her correspondence in the "Martin—Correspondence" file, then note this on your log, showing "SEE FILE Martin—Correspondence." The correspondence card then enables you to note quickly where the correspondence is located. If you establish this index of correspondence cards or electronic log early on in your search, you may avoid the frustration of trying to remember which file holds Mary Adams' letters and emails.

You can easily establish a correspondence log similar to the card shown in the figure. Investigate the options—consider a Microsoft Excel spreadsheet or personal information management software such as TreePad. You can even set one up using your word processing program, such as Microsoft Word. If using the latter, insert the correspondent's name alphabetically. The advantage of having the information on your computer is that you can use the program's **Find** or **Search** command, depending upon the program you are using, to locate those who are tracing specific branches of the family. You can find quickly all those tracing the "John and Mary (Smith) Jordan" branch.

CORRESPONDENCE CARD

Name

Address

City and State   Zip

Spouse

Family Traced

County Traced

Phone   Work Phone

Fax   email

Comments

SEE FILE: Updated:

Here's a sample of a correspondence card.

TREE TIPS

Don't be discouraged if you don't have a computer. Standard 4 × 6 cards will do for your correspondence index. Use a standard format for organizing the information on the card, and maintain the cards alphabetically by the name of the correspondent.

Using Computers to Cut Down on Paper

At first you'll keep every scrap of paper. You photocopy a marriage record from a book and photocopy the title page; now you have two pieces of paper in the file. Multiply this many times over, and you'll be overflowing with paper unless you find a way to reduce the stream. The problem isn't limited to providing the space needed to store all the paper. Consider that some day others will use the records. Unless your files are condensed sufficiently, others won't want to store them. Even repositories will be reluctant to accept them if they're in complete disorder or are too voluminous.

The Computer Becomes the Filing Cabinet

File on the computer in the same way you file in file drawers. Set up computer files by family name, state, county, subject, and correspondence, just as suggested for your paper notes. As you have time, take some of your paper notes, copy them to the computer, proofread them carefully, and then discard the paper copies. There may be certain copies you want to keep, but make your selections carefully. Always keep the following:

- Copies of Bible records

- Family notes (written in an ancestor's hand)

- Photocopies of original documents

- Birth, marriage, and death records

- Obituaries and other newspaper notices

Keep the items that won't be replaced easily, or that you may need to prove your documentation in the future. Items you can discard (after entering the data onto your computer) are items from published books and notes that can be replaced if necessary. If you transfer a marriage record from a book onto your computer and include the full bibliographic citation, you can find the book easily again, should you need to do so. Always be sure to proofread what you've entered before discarding the paperwork.

TREE TIPS

A good rule of thumb is to transfer to disc all items from books that you can access easily again. If the book is rare, the copy is from an original, or the text is lengthy, keep it. If the item is more than a paragraph or so, consider scanning it to your computer. Scanning can be a tremendous timesaver.

Naming the Computer Files

As you establish and name your computer files, be sure to keep some sort of master index to all your genealogy file names. You can then consult this master list to see what files you've established. When naming your computer files, use some designation for each family, and start each file name with that designation. For example, in the following, each of the Martin files starts with the name "Martin:"

INDEX TO GENEALOGY FILES

File Name

Martin: Alabama

Martin: Alabama, Limestone Co.

Martin: Bible Records

Martin: Correspondence A–M

Martin: Correspondence N–Z

Martin: Vital Records

You may wish to name your files in some other descriptive fashion. Anything will do, as long as you keep an index to them. You won't always remember their names. It's best to create a different electronic subdirectory for each family. You may name a directory (folder) "Martin Family" and then create all the Martin subdirectories under it.

Correspondence Computerized

At first you save every letter and a copy of every response. That works until you realize you have 10 thick files of correspondence. To go back and try to reduce the paper is time consuming. Start now, in the beginning, establishing some firm rules for saving those mail and email exchanges.

There are many ways to store your computer files. The one that works for me is to create correspondence files in the same manner as you would if you were using file folders. If you start one for correspondence A–M, it will hold the letters and responses to all correspondence with anyone of the surname A through M. If storing the summary on the computer, however, take a few minutes after you answer the letter to transfer the main points to the computer file. For example, you received a letter from Sandra Williams, in which she sent some information on the Martin family and asked for assistance. Examining the letter, you may summarize it on the computer as shown.

WILLIAMS, Sandra, 1111 NoWhere Street, Anywhere, U.S.A. 12345. Wrote 17 April 172007; says great-great-grandmother was Agatha Roberts who married Jonathan Martin on 3 April 1863, in Montgomery Co., Va. She thinks Jonathan was son of Roger Martin listed in tax records of the county in 1855 but is uncertain. This couple had: Jeffry b. ca 1866, Martha b. ca 1868, Mary b. ca 1870 (m. John Webster), and Joseph b. ca 1875. They're in 1870 Montgomery Co. census record (she doesn't give citation; wants to know if I have further info). Wrote to her 20 April 202007: told her I'm tracing a Joseph Martin who was born 4 April 1874, in Virginia by his tombstone in Hamilton Co., Ohio. I don't know if he's the same person, but I note that my Joseph named a son Roger Martin. Asked her for the listing she has in 1870 and sent her the 1860—70—80 census records of my Joseph Martin. Suggested we check deed records.

See MARTIN File.

Updated: 20 July 2011

Now you can discard the three-page letter she sent and your two-page response because you've recorded the essential details. If the correspondent writes again, add comments to the synopsis. The last notation at the bottom tells you when you last updated the contents of this summary.

TREE TIPS

If you started your genealogical pursuits by keeping all correspondence, set aside at least a half hour a day to transfer some of your material to the computer. It will go slowly, but doing it daily (in a dose small enough to avoid getting too bored) will enable you to complete it. You'll be encouraged to continue as you see the empty space in your filing cabinet grow, and the usability of the data increase.

You now have the notes in a handy form on the computer and you can search them with your program's **Search** or **Find** command. When someone writes three years from now inquiring about a Roger Martin, you can open your file of correspondence summaries, enter his name into a search, and presto! In a few seconds you've identified all correspondents with whom you exchanged data on him. Condensing your correspondents' letters onto the computer will take only a few minutes if you do it at the time the mail comes in or goes out. Another important consideration is that doing this will make your collection easily usable by others who come after you. Otherwise, they may throw up their hands and toss everything out.

TREE TIPS

Once or twice a year create a "Save This" CD-ROM or DVD disc or flash drive for all the files you wish your family to save. Do this in addition to the daily or weekly backup you make routinely of your computer files for your own use. This extra "Save This" disc contains specific files you want your family to retain. Put some thought into what your family should save for your future generations. Avert the disaster of having your years of research destroyed because no one knew what to save. (Be sure to store a copy in a safe deposit box or with an out-of-state relative.)

Mail Files on Your Computer

If you want to keep a full copy of every letter you write to others, you can set up a genealogy mail file on your computer; title it "Mail Genealogy 2012." Start a new file each year, but retain the old. In this mail file, keep a copy of your letters by simply adding each new letter to the end so the letters appear in chronological order. "But how will I find anything?" you ask. Easy. Want to find the letter to Jonathan Latimer you wrote in 2008 inquiring about his great-grandfather? Go to your "Genealogy Mail 2008" folder, use the **Find** or **Search** command, and enter "Latimer." The inquiry for which you're searching will surface quickly. You want to find the letter you wrote inquiring about the Oak Lawn Cemetery, but you don't remember to whom you wrote? Enter "Oak Lawn," click "Find," and there it is. Imagine what you would have gone through sorting paper letters by hand, trying to uncover it.

I don't store letterheads with my copy of the letters. I have several letterheads set up as templates on the computer—business, personal, spouse and I, mine alone, and so on. I start by writing the letter on the appropriate letterhead, and print the letter. Then I cut and paste the letter (not the letterhead) into the end of my mail folder, leaving the letterhead intact and ready for my next letter. Doing this can help to conserve storage space on your computer. If you have plenty of space, you can save the entire letter—including the letterhead—if you wish.

Commercial Products

The filing system suggested in this chapter enables you to easily expand your system with minimal cost to you. However, you may prefer to use one of the commercial products designed to help you store your records in binders or on the computer. Examine genealogical periodicals for advertisements. One popular program is Clooz, an electronic filing cabinet designed for genealogical records. It doesn't just record the results of your search; it's also a database for systematically organizing and storing all the clues you find. Go to clooz.com to determine whether it's right for you.

No matter what system you use, establish it now. Staying organized from the beginning will allow you more time to spend on what you enjoy most: the search for those elusive ancestors.

The Least You Need to Know

- Creating specific files for paper makes for easy retrieval.
- Correspondence index cards or an electronic log enable you to readily find letters and emails by directing you to the file in which you've stored them.
- Condensing correspondence onto the computer has several long-range benefits.
- Creating a DVD, CD-ROM, or flash drive periodically with the specific genealogy files you want your family to retain will help ensure that your years of research aren't lost.

Keeping Records Right

In This Chapter

- Utilizing citations to prove your information
- Organizing your compilation using numbering systems
- Approximating dates to assist in planning the search

You identified crucial dates and locations for your family, and doing so wasn't easy. In fact, you had to do an exhaustive search to find the information. Now you enter it into your compilation, like the following:

> John Washington Jackson was born 10 April 1857, in St. Louis, St. Louis County, Missouri, and died 5 July 1935 in Springfield, Greene County, Missouri.

You look it over and wonder if you missed something. Indeed you did—the citation is missing! On which record did you base the birth date, and on which did you base the birth place? It is important to let others know where you found the information on which the facts are based. The weight of the evidence cannot be judged without knowing their basis.

Past published genealogies commonly lacked citations. Present-day standards require a *citation* for every fact. It's easy to underestimate the importance of the citation, but don't forget you may need a source again, too. You suddenly realize items you copied from a book of abstracts appear to have a child missing. Did you just omit it when you copied it? You want to recheck. With the source citation, you can relocate the book quickly through the library catalog. Learn the correct procedures for citation when you first start your search and use them diligently—doing so will ensure you won't be constantly frustrated later when you try to find important records again.

DEFINITION

To cite is to call attention to the source of your information, expressed in a **citation** that gives enough detail for others to find the source and to help them judge the validity of the information it provides depending on the type of source cited.

Getting Help with Citing

There are books you can use to learn how to cite properly. One of the best and most detailed on style is *The Chicago Manual of Style*. Look also for two excellent guidebooks by Elizabeth Shown Mills. One is titled *Evidence! Citation and Analysis for the Family Historian*. It illustrates, simply and effectively, the manner in which almost every type of source encountered by genealogists should be cited. This includes books, magazine articles, journals, microfilm, manuscripts, CD-ROMs, and websites. A far more ambitious project—almost 900 pages!—is Shown Mill's *Evidence Explained: Citing History Source from Artifacts to Cyberspace*. Almost every type of record a genealogist would encounter is discussed in this manual.

TREE TIPS

The Chicago Manual of Style not only includes extensive information on preparing citations and referencing, but also other useful topics such as manuscript preparation and copy editing, indexing, and book design.

If you keep in mind that the purpose of a citation is to enable the reader to readily find the same source, it'll be easier to remember which items should be included. Basically, the citation for a book must include the complete title of the book, the full name of the author, the city and state where it was published, the name of the publisher, the year the source was published, and the edition of the source (that is, 1st ed., 2nd ed. revised, reprint, and so on). Citations for articles, microfilm, CD-ROMs, and other sources have specific rules on what to include so you or others can access the source. Don't omit any of the components.

TREE TIPS

Some researchers photocopy the title page of the book to prepare the citation. Though this may be helpful, it adds expense—and it adds "paper" to your files. Learn early the correct citation techniques, and write the details on the photocopy. Check also the reverse of the title pages of books; some of the needed information for citing may be there.

Cite personal documents, such as letters and photographs. The citation for a letter Aunt Martha wrote to Cousin Jim in 1910 giving the family history should be cited by the guidelines given in the manuals, as should an oral interview, tape recordings, a video, or handwritten notes. For Bibles, besides the publication date, note whether your information is from the original Bible, a photocopy, a handwritten transcription, or a typed copy. Doing so will assist in evaluating whether errors may have crept in while copying or transcribing. (For more detailed examples and illustrations, refer to the aforementioned manuals.)

You may think, "But I am only doing this for fun." Your diligent search for your ancestors, if not cited properly, could someday puzzle and even mislead descendants who'll wonder upon what the statements were based. Your record may be correct, but they won't know unless they know where you got your information and the type of source from which you took that information.

Undocumented data is an immense problem for genealogists. In the past, researchers made little effort to include citations. Today there's increased interest in restudying earlier records to add details and to thoroughly document compilations to correct prior errors and omissions. This applies not only to dates, but also to locations, parentage, and other important details.

TREE TIPS

The necessity of adding citations isn't a reflection on your ability. You may use a record that appears correct when you locate it; another record you didn't even know about that conflicts may be found later. Knowing the source of each statement will help you to judge the merits of each.

Family Group Sheets Are Convenient, But ...

Family group sheets, those forms leaving room for the names and statistics of an individual family, are popular because they're a convenient form of entering data. Seldom do the forms have sufficient room to enter the sources, though. Computer programs are now usually able to handle the problem by allowing the user to include text notes. Family group sheets prepared in earlier times, without the benefit of a computer, encouraged the researchers to enter dates and locations without documentation. Genealogists who attempted documentation normally wrote on the back of the form, which was often omitted when the documents were photocopied for others.

Putting It at the Bottom or at the End?

If you're using a standard word processor for your compilation, you can add your citations in one of three ways:

- In-text citation
- Footnotes
- Endnotes

The in-text citation adds the information immediately following the fact:

> Mary Morgan was born 3 April 1846 [Family Bible, adding the full citation here]. She died 8 March 1895 [Death Certificate: adding the full citation here].

As you can see, the text is broken up considerably with each citation. This method has fallen out of favor because of the choppiness; it's best if you disregard this method. There are two favored systems: a footnote system (in which you reference citations that appear at the bottom of each page) and an endnote system (in which you reference citations that appear at the end of the chapter or compilation).

Because many word-processing programs can handle footnotes, they're easy to implement. In this system, you add a superscript number to the statement in the body, which keys it to the footnote that holds the citation, which appears at the bottom of the page. Sophisticated programs renumber footnotes automatically as you add or delete them. The advantage of footnotes is that the citation remains on the same page as the text to which it refers.

In endnotes, the superscript number refers to a citation or notation that appears either at the end of the chapter or the end of the compilation. Though permissible to use this system, readers may find it annoying to constantly flip pages.

Use the recommended citation system that suits you; the important things are that the citations are there, correlating directly to the statement you're making, and that your citations are consistent in style. It's not enough to add just a list of "Sources" to the end of the compilation. The person using the data must know which source specifically goes with which fact.

TREE TIPS

Whenever examining a family sheet sent by others, note whether it's cited fully. If not, use it only for clues. The information may be correct, but you'll have to check each bit of information before accepting it as fact.

Keeping Track by Numbers

When you start using published genealogies in the library, you'll be stumped many times by complicated numbering systems. When faced with a massive list of descendants in a family, compilers often devise their own system to keep track. Later, they decide to print it for the family. Being familiar with their own style, they resist any thought of transferring it to a uniformly accepted system. The increased use of computers seems to have encouraged many to create their own, forcing the reader to devote considerable time to understanding a complicated or offbeat system.

An excellent publication by Joan Ferris Curran, Madilyn Crane, and Dr. John H. Wray, *Numbering Your Genealogy: Basic Systems, Complex Families and International Kin*, can assist you in understanding the numbering systems widely accepted by genealogists.

TREE TIPS

Even if you never publish your material, you'll likely distribute copies to your family and relatives to read. A good numbering system enables them to understand what you've compiled.

Descending Genealogies

The two most widely used and recommended systems are the Register System (developed by the New England Historic Genealogical Society for their NEHGS *Register*), and the National Genealogical Society Quarterly (NGSQ) Numbering System (formerly known as Modified Register System) used by the *National Genealogical Society Quarterly*. These systems apply to descending genealogies. The Register System normally starts with the immigrant, but if the immigrant's identity is unknown, it can start from the earliest-known ancestor of the family. He or she is assigned an Arabic numeral 1. The children of number 1 are all given Roman numerals, in order of birth. In the Register System, those for whom there's sufficient information to carry forward include the next Arabic numeral in front of the Roman numeral, alerting the reader that more will be found in the section corresponding to that Arabic number.

In the following figure, there's no need to carry child "i." to her own section; the researcher doesn't know enough about her to warrant carrying her forward in the compilation. Mary, child "iii." died young (d.y.) and isn't carried forward either. The researcher knows more information about Jonathan, so Jonathan received Arabic numeral 2, notifying the reader there's more information on him in section 2. There are more details available for Ezekiel, so he received Arabic numeral 3. Those Arabic numerals precede the sections in which the researcher has presented more details on Jonathan and Ezekiel.

---

1. John[1] Smith was born 3 April 1788, etc. [include here rest of information on him, events of life, marriage, documents, etc.]

   Children; surname SMITH:

       i.   Mercy b. 5 Apr. 1808, Richmond, Va., further unknown

   2. ii.   Jonathan b. 18 Mar. 1810, Richmond, Va.

       iii. Mary b. 3 Aug. 1813, Richmond, Va., d.y. 4 May 1814

   3. iv. Ezekiel b. 22 Dec. 1816, Alexandria, Va.

2. Jonathan[2] Smith (John[1]) was born 18 March 1810, Richmond, Virginia, etc.

3. Ezekiel[2] Smith (John[1]) was born 22 December 1816, Alexandria, Virginia, etc.

The NEHGS Register system of numbering.

Note that generational numbers are also included in the figure. John[1] Smith indicates that he's either the immigrant, or the first identified progenitor in this family. The children who are carried forward are also shown with a generational number in superscript, followed by a "runner" that shows their line back to the earliest known progenitor.

Let's say you've researched Abraham Smith of the fifth generation. He's carried forward to his own section. The generational runner would be something like this: Abraham[5] Smith (George[4], William[3], Jonathan[2], John[1]). It's easy to distinguish these generational numbers from the compilation's overall numbering, and from any citation numbers you insert because the superscript generational numbers are inserted after the first name of the individual, whereas superscript numbers for citations follow the surname or sentence. Thus, John[2] Smith is referring to John Smith of the second generation. John Smith[2] indicates a citation reference.

TREE TIPS

If you're using a computer, use a somewhat larger typeface for the numbers that precede the sections, and insert them in boldface (as illustrated in the preceding figure). This large, bold type will help distinguish section numbers preceding a specific section from the reference numbers preceding the Roman numerals in the children's listing.

The NGSQ System

The NGSQ System is based on the Register System, but it assigns an Arabic numeral to each child. The plus sign (+) in front of the Arabic numeral designates whether the child is carried forward.

As seen in the figure below, the plus sign readily identifies the sections that hold further details when carried forward. The sections that are carried forward are preceded by the Arabic numeral assigned to that individual (in large, bold typeface if using a computer).

---

1. John[1] Smith was born 3 April 1788, etc.

Known children of John Smith and wife Mary Martin were as follows:

    2.   i. Mercy b. 5 Apr. 1808, Richmond, Va., further unknown

+   **3.**   ii. Jonathan b. 18 Mar. 1810, Richmond, Va.

    4.   iii. Mary b. 3 Aug. 1813, Alexandria, Va., d.y. 4 May 1814

+   **5.**   iv. Ezekiel b. 22 Dec. 1816, Alexandria, Va.

3. Jonathan[2] Smith (John[1]) was born 18 March 1810, Richmond, Virginia, etc.

5. Ezekiel[2] Smith (John[1]) was born 22 December 1816, Alexandria, Virginia, etc.

The NGSQ System.

Each of these systems has devotees. Those using the Register System find it simple and neat in appearance, but most admit that it has a disadvantage: when the compiler must add newly found information on a child whose data was unknown previously, creating a new section to carry forward that child's data requires complete renumbering.

Those who prefer the NGSQ System like the advantage of having an Arabic numeral already assigned to each child. If more information is located on a child, it's simple to insert a plus sign before the Arabic numeral, and insert the section in its proper place. Even those using the NGSQ System have to renumber, however, if they discover the couple had more or fewer children than was originally thought when the researcher first numbered the compilation, but most computer programs automatically renumber the NGSG system for you.

Before They Came

Because the immigrant is numbered with an Arabic 1, you're likely wondering how to denote the generations before the immigrant's arrival to the United States. You're tracing a family from England and have information on earlier generations. You would handle these by assigning the closest pre-American generation as A, the next as B, and so on. If the father of immigrant John Jackson was George Jackson, and his grandfather was William Jackson, this would appear John[1] Jackson (George[A], William[B]). This immediately conveys that John is the immigrant, his father is George, and his grandfather is William.

The Register System and the NGSQ System lay out the family generation by generation. The systems are easy and will soon become second nature to you. You can study further details and illustrations in *Numbering Your Genealogy* mentioned previously in this chapter. Also see *The BCG Genealogical Standards Manual* for utilizing numbering systems.

Approximating Dates

If an exact date is unknown, estimate the dates whenever possible rather than leaving them blank. In this section, we discuss a few of the many ways in which you can accomplish this.

By Date of Will

If John Morgan made his will on 17 April 1853, and the will was presented in court for probate on 15 July 1853, you know John Morgan died between 17 April 1853 and 15 July 1853. It's best not to say simply "died 1853." Rather, state more specifically that he "died between 17 April 1853 (date of will), and 15 July 1853 (date will proved)." Be as precise as you can in presenting information if you have the records to do so. It will prevent confusion when someone finds an entry that indicates John Morgan sold a piece of land on 12 February 1853. If you had entered the date only as "ca. 1853" (about 1853), the researcher may have been puzzled; if you entered the death as being between the date of the will and the date it was proved, those using your records will realize he was living until at least 17 April.

Rule of Thumb

Studies of countless families living in the seventeenth and eighteenth centuries (and earlier) have determined that a good rule of thumb is that children were born two years apart. When you examine the lists of children of your earlier relatives, you may find gaps, but these often can be accounted for by the death of a child. You know the couple married in 1816, their first child Amy was born in 1818, Timothy was born in 1822, and their young child Jonathan was born in 1824. When you discover a child Joseph, you can estimate he was born ca. 1820. Don't state this date as absolute fact; instead, include "ca." (*circa*, about) to indicate that it's an approximated date.

Similarly, if you know the marriage date and don't have the birth dates of any of the children, you can approximate their births based on the marriage date. If John Taylor was married in 1838, and he had three sons and a younger daughter, you can estimate the sons were born ca. 1840, 1842, and 1844, and the daughter ca. 1846, two years apart.

You can also use census records to provide an approximated date. If the age of John is 15 in the 1850 census, 22 in the 1860 census, and 37 in the 1870 census, you can approximate he was born ca. 1833–1838.

TREE TIPS

In the example I've given, we not only used the rule of thumb, but also combined that with the fact that Jonathan was the youngest and that the couple married in 1816. It's often possible to glean clues to an approximate date from several related records.

Choosing Their Own Guardian

Minors had guardians appointed by the court if they had any property due to them. Most states allowed those minors, upon reaching the age of 14, to choose their own guardians. If your ancestor is noted in court records as making that choice, you can estimate he or she was between the ages of 14 and 21. (By the latter age, a guardian wouldn't be necessary if the person was deemed competent.)

TREE TIPS

As you progress in your genealogical pursuits, you'll turn to the laws to determine the legal age for certain matters to best evaluate the ages involved. Remember that states differed and laws changed.

Of Age to Marry?

Sometimes you may find records of those who misrepresented their ages to get a marriage license; nonetheless, the marriage license can be a gauge for determining age. If the couple applied for the license without the consent of a parent or guardian, assume they were of age unless other records prove otherwise. The legal age varied by location and time period, but usually the groom had to be 21 years old and the bride 18 years old (legal ages were lower in some areas, especially for brides).

Why Bother with Approximate Dates?

The purpose of establishing an approximate date rather than leaving it blank is to better determine the records you need to examine. If you know only that your ancestor William Carr was born in Tennessee, then you have nothing on which to determine a course of action. If you know he was born ca. 1791 in Tennessee, you see by his age he may have served in the War of 1812 and you can investigate the possibility of military service. Or, knowing he was born ca. 1791, you can examine the 1820 census records of Tennessee to see if he was listed as the head of a household because he was about 29 at the time of that census. Other ideas will come to you, too, after you've estimated his birth date.

The Little Children

Always include in your compilations the little children who were stillborn or who died before maturity. They, too, were important in the overall genealogy. They may carry a name that's significant—perhaps named after the mother, grandmother, or another family member. That name may be the clue that triggers a breakthrough. Their ages are important; they can explain gaps in the family groupings. Their burials are important; they may lead to family plots. Lacking any other reason, those children were precious to their families. If you don't remember them, no one will. Even their causes of death can explain a mystery; if the child died of cholera, you may understand why three of the family members disappeared from the records that year.

Genealogy should be fun, and it is. Developing sound techniques doesn't diminish that fun—it adds to it. Taking the time to do things right from the beginning will save you many headaches later. It will also preserve your information in such a way that you'll be proud to share the results with others. You'll know you've produced a family history documented to the best of your ability, and that it will be easily understood by those who read it.

The Least You Need to Know

- Citations are important; they enable others to locate the sources you used.
- Knowing the source helps in evaluating the trustworthiness of the information.
- Numbering systems should be standard.
- Approximating dates will assist in pointing you in the right direction for the search.

Gaining Historical Perspective

In This Chapter

- Gathering more than the bare facts
- Analyzing and interpreting the documents
- Interviewing the oldest relatives
- Writing The Life Story: a practical example

You've been gathering the bare facts about your ancestors: their names, dates, and locations. So far, so good. At this point, though, they're just shadowy figures. Who were these people whose genes you share? You can connect with them and make them come alive, if only on the pages of your family history. How can you know what life was like for them? Make the records talk to you. Add historical background and visualization so your ancestors become more real to you. You enhance your family history by looking deeper into the records, scouring historical material for its relevance to your ancestors, and gathering stories and artifacts from relatives.

Taking a Fresh Look at What You've Found

By analyzing the records, you'll find the information to flesh out your ancestors. You dutifully wrote down (preserving all spelling, of course!) the items listed in the inventory taken when your ancestor's estate was probated, and you filed it away. Take it out and look at it again. What does it say to you? Even the arrangement of the inventory can give you clues. Think about how you inventory your possessions for insurance purposes. You start in one room and progress through the house. That's the way most people take inventories. Stand in the doorway with the appraiser of Elias Drollinger's possessions and look around the room with him. (Note the spelling and capitalization in the following is as it was in the original.)

two beds and bedsteads

four sheets four pillows six blanketts

two double coverlets

one Bureau and stand

table

chest

two sets bed curtains

stove

one set chairs

rocking chair

little oven

one set silver teaspoons

one set silver tablespoons

molases can

sugar bowl

tea canister

pitcher set

red plates

one saw

two trunks

looking glass

one wash tub

wash board

sugar tub

lard tub

one flat Iron

one par hand irons

one bible and 10 Books

one cow

The possessions in the preceding list, meager by our standards, probably describe a house with two rooms. One room had two beds complete with bedding and curtains for privacy. The other room was the all-purpose kitchen and family-, dining-, living-, laundry-room. The looking glass hangs on a nail over the washtub that was also used for bathing. Picture the saw, leaning against the wall, perhaps next to the stove, and the table set with red plates and silver spoons. Where are the books and Bible? (And who got the Bible with the family records?) Someone in the family was literate, hence the books. Outside is one cow, probably kept for milk, cream, and butter. No implements suggest an occupation. Was the individual elderly and perhaps retired?

TREE TIPS

Use your imagination to bring your ancestors to life. Just be careful not to blur the line between nonfiction and fiction. You aren't writing a novel.

Expanding Your Research

Look for records that give you data with which to work. Comparing the New York state census records of 1855 and 1875, you can learn a great deal about your farming ancestor. For example, suppose in 1855, an enumerator in Onondaga County mentioned yields were down one half to one third because of drought and wind, yet your ancestor's crop yields were much better than those of the neighboring farmers. From that you know he either was a better farmer or had better land. You learn he harvested 100 pounds of honey; he probably kept bees to pollinate the fruit trees, because he also had 100 bushels of apples and 20 gallons of cider. In 1875, his crop land allocations were quite different. By this time, he was 75 years old, so it's conceivable he was slowing down. Yet 15 years later, a newspaper article remarks that at 90 he's the oldest native-born resident of the county and "is an excellent farmer, though nearly blind." These small details enhance the picture you're developing.

Does the deed you found tell what your great-grandmother paid for her house? When she sold it, did she make money on it? Use one of the inflation calculators such as westegg.com/inflation to relate those amounts to today's monetary values. What do newspaper ads from your ancestor's time tell you it cost him to purchase necessities he couldn't grow or make? On what was he taxed? The minutia of your ancestor's life is what makes her more than just a name on a pedigree chart. Would you want less for yourself?

Court Records Tell Many Secrets

Your ancestors' court appearances give you more insight. They appear in court for "unlawful gaming" or because a speculator wants to foreclose on their land. They want water rights, or they're evicting a tenant. There may be statements in their own words.

Divorce records can be quite telling. He accuses her of infidelity; she accuses him of drinking too much and hitting her. After the divorce is granted, they continue to appear in court, wrangling over their young daughter. The mother says the father doesn't meet the little girl's train; the father says the mother doesn't send her dressed warmly enough. Reading this, you understand a little more about the bitterness in the family and how the family became estranged.

TREE TIPS

Deed books often contain more than land records. They may include the record of an adoption, an apprenticeship, or a livestock brand. Perhaps a bill of sale for "one yoke of oxen, one red and one brindle color, I call one Bill, the other John."

Treasuring Your Ancestors' Enemies!

Disputes create records and provide glimpses of your ancestor you might not otherwise have. You never know what you'll find. In a deed book written in 1805 in Bourbon County, Kentucky, Peter Sap releases Thomas Glass from further liability for having cut or bitten off the top part of his ear in "an affray."

You may find something similar to my own discovery: my great-grandpa's Civil War pension was challenged by a disgruntled townsman who wrote his congressman, saying, "after coming home they all say he been an auctioneer, if a man is very deaf it does seem to me that a deaf man would make a very poor auctioneer." As a result of this complaint, there was an investigation, including a physical examination that showed the old man to be quite frail, and testimony from numerous townsfolk about Great-Grandpa's fine character. Your ancestor's enemies can be just as important as their friends in generating records for you to find!

Unexpected Finds

There's much that can be said for examining the original records even if you have seen them on film, but especially so if you've seen only the abstracts. Perhaps the record has a notation that the abstracter didn't include. In Wabash County, Indiana, Marriage Book 6, there are three handwritten notes pasted on the inside front cover. The notes were written to the County Clerk by three fathers telling him not to issue marriage licenses to their children. My favorite appears in the following:

> Wabash County, Ind. Jan. 11, 1867
>
> Bro. E. Hackleman Dear Sir do not issue a marriage license to Marshall Murray & Mary E. Shortridge without a certificate from me. Mary is underage & I am not being treated as respectfully as could be desired if everything goes off right I will give them a certificate.
> Respectfully   L. Shortridge [signed]

You may be lucky enough to find a record that's annotated and gives you totally unexpected information: the census taker who marks a name with an asterisk and at the bottom of the page writes "died of drink," or the assessor who adds comments beside the names on his list: "bankrupt," "left for Kansas," or, my favorite, "not dead, but asleep."

By scrolling to the end of the Sevier County, Utah, 1870, census roll, I found this notation: "The foregoing number of houses were abandoned by the Whites during the late Indian War of 1865—'6 —'7." If your ancestors were in this area then, this could well be relevant to their lives, and you'll want to follow up by reading about the events that took place there.

LINEAGE LESSONS

Visiting living-history sites is the best way to gain insight into the lives of your ancestors. A good substitute, however, is a vicarious trip via the web. For a fabulous array of photographs of Plimouth Plantation, Sturbridge Village, and Williamsburg, go to the website of Galen Frysinger (galenfrysinger.com/ plimouth_plantation_mass.htm).

Looking Beyond the Records

Follow every opportunity to learn about the times of your ancestors. All over the United States there are towns and museums creating living-history celebrations. Colonial Williamsburg in Virginia, Sturbridge Village and Plimouth Village in Massachusetts, Skagway in Alaska, and Columbia in California, are some of the large ones, but there are many small projects, too. For now, attending these is the only kind of time travel we can know. If your ancestor was a blacksmith, visit the recreated blacksmith's shop. Observing the activity will make you realize how much physical strength your ancestor had to possess to pursue this occupation. Visit the kitchen garden where your ancestor grew things for the table and the medicine cabinet. Families had to be self-reliant, and mothers learned the benefits of many herbs for treating illnesses.

Board the full-scale ship replicas and consider how you would survive an ocean crossing. Visit a colonial farm, a plantation, or a Victorian home. The Wild West really was wild. Look for reenactments of gunfights and train robberies. Tour a coal mine or a gold mine.

Libraries, museums, and historical sites continue to digitize their collections. The millions of documents, photographs, films, and recordings online can distract you for hours. The Library of Congress's online archive, known as the American Memory (memory.loc.gov/ammem), astounds with its breadth. The mere thought of searching for something specific to your ancestors or to their experiences is dizzying. But the searches are well worth the effort. An obituary states Frederick Harkrider was a U.S. guager [*sic*]. Learning that a guager was a revenuer—specifically, a fed who measured or gauged the whiskey for tax assessment—I searched the American Memory for more information. The sheet-music collection includes an 1883 ditty "The Wine Guager," by E. W. Foster, that opens with …

> "A wine guager down in its vault so cold,
>
> Thus sang to himself a song.
>
> 'Oh what care I for place or gold,
>
> 'Tis wine makes me happy and strong.'"

That definitely adds color and perspective to the family history!

DEFINITION

Sic means "thus" or "in this manner," and is usually italicized and enclosed in square brackets to indicate a misspelling or an unusual spelling, or to otherwise call attention to something that could be perceived as an error.

Tornado Watch

Weather is a factor in our everyday lives, but it loomed even larger in the lives of our ancestors. Their crops were dependent on the rain. They learned how to prepare for the usual occurrences of harsh winters and blistering summers. It was the unusual that dealt them severe setbacks: the tornado, hailstorm, or flood. Families in precarious financial situations were pushed over the edge by bad weather. This can affect your research and may explain why families pulled up stakes and left an area.

Reading about the weather can help you interpret the records or support a family tradition. The family says Fred drowned in the creek in 1892; you're unable to find an obituary or death record. A local newspaper from that week tells of heavy rains and flash floods in the creeks, which lends some credence to the story.

Being a Student of Local and Family History

As you try to complete the pictures of your ancestors' lives, steep yourself in the history of the period and locality; there may be clues there. Reading about the coal mine disaster or the opening of the canal, you may finally understand why your family left the area. The newspaper article in the weekly paper extolling the virtues of Nebraska may explain why they left the fertile land of central Illinois and headed west.

Our ancestors were often passionate about their religions, and religious upheaval was common. Arguments over doctrine among the parishioners often caused a dissident group to start another place of worship—down the road, in another county, or in another state.

Letters or diaries that pertain to your ancestors' experiences may give you insight. Seek them in books, manuscript collections, and online at university and state libraries or at sites such as eyewitnesstohistory.com. Even entering terms such as "witness to history" followed by the subject, such as "holocaust," will produce a list of websites.

PEDIGREE PITFALLS

Be sure historical events are relevant to your ancestors before including them in your narrative. The lives of famous people probably touched your ancestors no more than they do yours. But legislative action such as the Homestead Act or events such as the Gold Rush may well be important.

Illustrating Your Story

Don't be satisfied with just portrait photographs of your ancestors. Collect photographs and maps of the local area. The National Archives in Washington, D.C., has millions of still photographs with an incredible range of topics. There are photographs of military units, photographs documenting the terrible conditions in the Dust Bowl, and photographs of federal projects. The grandeur of national parks and the bleakness of tenements are preserved. On a less sweeping scale, local museums and historical societies tell the stories of their counties. It's there you'll find the picture of Grandpa self-confidently astride his horse and Grandma posing with the students she taught in the one-room school.

Visit online photographic collections such as photoswest.org, a part of the Western History Collection at the Denver Public Library.

If you know the ship that brought your ancestor to the United States, look for a picture of it or a similar ship. Maritime museums are a good source for sketches and photographs of old ships. Check theshipslist.com, which has not only illustrations of ships, but also descriptions of their voyages, travelers' diaries, and other resources.

Historical maps and atlases will help you locate the exact spot of the family farm. The atlases may have drawings of local sights, such as the livery stable in the following figure. The caption under the sketch says, "The only first class livery in Morris. Ladies & gent's driving horses a specialty. Particular attention paid to funerals. Good accommodations for farmers." Your ancestors in Grundy County may have patronized this place, but even if they didn't, it was probably a part of the scenery of their everyday life.

Here, a livery stable is sketched in Atlas of Grundy County and the State of Illinois *(Chicago: Warners & Beers, 1874).*

Closer to Home

In addition to analyzing the records and reading history books, contact all the relatives you can find. Ask them to look in their attics, basements, and filing cabinets to see if they have some long-forgotten piece of family history. Carefully examine jewelry and pin-back buttons with letters or symbols that indicate membership in a fraternal organization. For help in identifying the organization, go to exonumia.com/art/society.htm for a comprehensive list of secret societies and fraternal orders.

This particular part of your genealogy research depends a great deal on serendipitous finds. From a family you only recently learned existed, you may receive a copy of a letter such as this one written on 15 December 1875, in Bear Grove, Iowa, from Alice Dunn to her cousin Lis Harcourt [spelling as in the original]:

Dear cousin I received your most welcome letter yesterday and was very glad to hear from you once more. It found us all well. I was at church when I got it. Our father got it from the office and brought it to church and gave it to me …

Well you said you expected my school was out. So it is but I have taken another one but for a longer term than that one. I have taken one for life this time. How is that for high. You know we promised to tell each other when we were going to get married. Well I wrote you a letter and told you to answer it right away and I would tell you something but I guess you did not get the letter from the tone of your letter. I was married on the 18th of November 1875 at home at seven 7 o'clock in the evening and also Willson was married the same day at twelve in the morning. He married a lady by the name of Miss Nora Mason and my name is Mrs. Parsons. I guess you know the rest of his name. I wish you could have been there. We had a real nice time. There was about 30 at the wedding. I will send you a piece of my wedding dresss and tell you how I made it. Well I trimmed the front breadth in knife pleatings and sheared ruffing and the rest of the skirt on down and put it on in box plaiting and put a mould over each plait and made it to trail and looped the skirt. The back bredth were just one yard and three quarters long. I got it in Desmoines city and my hat is white velvet trimmed in drab rep ribbon and drab plume and veil. My veil is one yard and a half long. Come over and see me. I am keeping house now. I live just one mile from fathers. I am going to look for you in one year from now and I want you to be sure and come. I expect it will be your weding tour. Be shure and write me when it is and I will come and then you can come on home with me and we will have some buckwheat cakes and beef juice and dried turkey and boiled frog and old rooster and row potatoes boiled [?] and stewed cake and fried biscuit and stewed onions and baked beets and roasted turnips and other things to numerous to mention. I will send you our picture as soon as we get some taken but do not wait for them but send your picture without delay. John and Willson are going to start for Marion Co this state next week and me and Ed are going to go with them and visit so no more at this time.

Please answer soon. Give my love to all enquiring friends and keep a portion for yourself. From your cousin Alice.

Here's a wonderful glimpse of life. The new bride excitedly tells her cousin about her wedding dress, and details the feast they'll have when the cousin visits. Notice, also, the genealogical information. Alice gives her husband's name and her brother's as well as that of his bride, and the date and locations for both weddings.

The Art of the Interview

You want to conduct interviews with everyone still living who knew your relatives and remembers the stories that were handed down. Don't delay. Deaths and illnesses may put their recollections and insights forever beyond your reach. Glean all you can from still-living family members and their associates. Seek out the stories of their youth.

You want a general feel for the events and traditions of the family, and you want to know about the individuals; design your questions with that in mind. You have to be more explicit than "Tell me about yourself and tell me about the family and friends."

The "things" of our ancestors are catalysts for stories. When family treasures are handed down, they're accompanied by tidbits such as "Mama always made potato salad in this bowl" or "My mother won these pearls at the raffle at St. Joseph's, and she wasn't even Catholic." Often, quilts are made from pieces of clothing the family wore. (If the stories are from a time beyond the memory of anyone living, be careful in accepting them as completely accurate. The story probably has a kernel of truth, but may have been altered from years of telling.)

Appendix C has suggestions for interview questions. For more help in devising your list of questions, go to cyndislist.com and scroll to "Oral History & Interviews." You'll find websites that discuss the art of oral interviews, guides for interviewing, and tips and tools for oral interviews.

TREE TIPS

Always have a list of interview questions, starting with the things you most want to know. But be flexible, ready to go in a new direction as the reminiscing progresses. Make the interview a pleasant conversation, not an interrogation.

We can't know our parents as vigorous young adults the way their contemporaries did, but we can learn about them by interviewing their friends. You'll be amazed at the stories their friends share—the time your mother borrowed the dishes to impress your father's wealthy aunt who was coming to dinner and then forgot to serve the rolls; how your father grew beautiful flowers, but he showed his practical side by planting the daffodils marching along the garden border rather than flamboyantly scattering the bulbs.

TREE TIPS

Older people often feel self-conscious talking into a tape recorder. Make it unobtrusive so they forget it's there. I had great success in taking my mother 2,000 miles to the town where she grew up. As we drove slowly around, the tape recorder on the dash recorded her excited recounting of past events and people.

Putting It into Practice

Let's consider an example of how you might direct your research to build on the facts you've accumulated, using the previous suggestions of contact with relatives, interpretation of records, and reading historical background.

> Anson Parmilee Stone, born 9 January 1815, in Oneida County, New York, married 14 October 1835, in Vernon, New York, to Cornelia Adams, daughter of Isaac Ward Adams and Eunice Webster. She was born in Vernon on 5 May 1812, and was a distant cousin of the two Presidents Adams. Her grandfather Abraham Webster was brother of Noah Webster, the lexicographer. Anson died 14 March 1852, in Ft. Atkinson, Jefferson County, Wisconsin, and is buried there at the Lake View Cemetery. Cornelia died 16 February 1882, in Ft. Atkinson, and is buried with him. They had five children, the three oldest born in New York in 1837, 1839, and 1842, and the two youngest born in Ft. Atkinson, Wisconsin, in 1845 and 1850.

The preceding contains the facts. But we want to know more. One of the first items that might interest us is the trip from New York to Wisconsin sometime between 1842 and 1845 (based on the ages of the children). What took them there? How did they travel?

Contacting and Interviewing Living Relatives

Following the suggested pattern of research, the researcher conducted interviews with all the grandchildren who were living when the researcher was completing the research in the 1960s. They were interviewed exhaustively about their recollections, and each scoured his or her own collection of family material. A handwritten paper was found in Anson's son Marsena's hand, stating that "Father went ahead out west, to grow up with the country, and he bought a farm, 200 acres, one half mile from Ft. Atkinson and built a small house and a large barn, split rails and fenced in the farm …." Anson sent word to his wife Cornelia to bring the family, and according to the same writer, they joined him in 1844, making the trip from New York to Milwaukee. Said Marsena, "[Father] sent for Mother, Emory, Newton and little me, we followed up on a sailing scooner across the lakes. Newt told me when I had grown up that we had a terrible squally time crossing the lakes and I squalled so loud and long that he and Em put their heads together to throw me over board, but Mother never gave them a chance. Theres no one that walks on earth like a mother …."

Use the events in the lives of your ancestors to trigger a search for interesting side-lights. The event mentioned in Marsena's letter led to a study of travel via the Great Lakes. Cornelia and her children took part in the great sailing era that began about 1834. The three boys, ages seven, five, and two, were no doubt enthralled with the billowing sails of the schooners crowding the Great Lakes, whereas Cornelia was preoccupied with trying to keep them from falling overboard.

There was no Internet research in the 1960s, but today you could add to your knowledge by examining sites returned from an online search for "Great Lakes schooners."

Checking Records for More Details

The events of Anson and Cornelia's lives were chronicled in the records. Census listings showed Anson was a farmer. His estate record was kept in county records. His will revealed he was in "a very infirm state of health" when he made it in 1852. He left his wife Cornelia in control until the oldest child became 21 years of age. He added that she was to treat each in a manner as "nearly equal as possible" but "always giving those that are the most needy and unfortunate the preference." He desired his estate be sold and the family maintained from the funds. He specified that funds were to be for support and education.

His obituary confirmed his parentage and his birth, and his age at death as 37. It added that at age 14 he joined the Methodist Episcopal Church, and served as class leader and steward most of the time until his death. He died of pulmonary consumption (tuberculosis).

Beginning to Know and Understand

Now we start to understand this family and the lives they led. We sense the struggle of the family: moving from New York to Wisconsin and then having the father die before reaching 40; the young mother, nursing her sick, contagious husband, then left alone to raise five children and keep the farm going. Researching the land sales, we find Cornelia sold the farm in pieces for their maintenance, following Anson's instructions. She kept some of the land, and was listed as a farm widow in 1860. In 1880, although no longer on the farm, Cornelia was still in Ft. Atkinson, living with her school-teacher daughter.

To continue the story of the family's life, the researcher followed the lives of the children, too. The son Marsena served in the Civil War, opening a new study area for details to add to his life story. The children began marrying, and some moved to Illinois, Missouri, and ultimately California, when the transcontinental train was connected. (The historical events surrounding the completion of the train route also provide fascinating details.) Another son became a doctor and went to Montana, while a third became a dentist and headed to Alaska. Cornelia did well, seeing that the children were educated. The family continued to make its way, and then adventure and the Far West beckoned the children, just as it had their parents.

Writing the Story

Writing your family history is more than a mere rearrangement of the family group sheet into a narrative paragraph, but you need not be a professional writer to complete the picture. If you have difficulty, write brief sections with the goal of piecing it all together. Try writing it as a letter to some other family member.

Never think your writing must be polished and of publishing caliber to be worthwhile. Think of how excited you are to find *anything* written by your ancestors. What your reader wants are the details that perhaps only you can supply. A bonus of writing about your ancestors is that you'll quickly see what you're missing and what events you want to investigate.

Your ancestors were ordinary people, making good and bad choices, raising wayward teenagers, and caring for elderly parents. These stories are about them—real people coping with real problems and successes. They're as much a thread in the fabric of history as the important personages in the history books.

TREE TIPS

Many books can give you ideas of how to craft the stories you're telling. One of the best is Lawrence Gouldrup's *Writing the Family Narrative*. He includes an annotated bibliography of books written about families. The books he lists are examples of various approaches.

The Least You Need to Know

- Reconstruct your ancestors' lives by interpreting their records.
- Look for records that amplify and expand the dry statistics you've accumulated.
- Use the vast resources of the Internet to find enhancements for the life stories you're writing.
- Read historical fiction and nonfiction about the people, places, and events of your ancestors' times.
- Interview your elderly relatives now.

Resolving Discrepancies

In This Chapter

- Addressing data conflicts
- Thinking creatively to find data
- Understanding customs that impact your research
- Applying calendar changes to your research
- Understanding evidence

As your research progresses, you may begin to notice information that doesn't agree. Dates don't seem to work out correctly or you seem to have two children in the family with the same name. You can't find someone you're sure was in a certain place. "What's happening here?" you may wonder.

To solve these mysteries and others, you need strategies to deal with conflicting data and a lack of information.

When Information Doesn't Agree

The census says your great-grandfather was 9 years old in 1870, but the 1900 census record says he was born in 1860. His tombstone says he was born 4 April 1861. The Bible record in his father's pension file has 7 April 1861. Which is correct?

First, recheck your notes and sources to be sure you didn't make a mistake. Then, analyze the information you've collected. Evaluate the sources from which you collected the information. Ask the following questions and apply the answers to your problem.

- Was the information from a published source or from an original document?

- If from a published source, what do you know about its reliability?

- Who supplied the information? Was there anything to be gained or lost by giving false information?

- Was the document created at the time of the event?

- Have you misinterpreted something because you're unfamiliar with a custom, an abbreviation, or the use of certain words?

TREE TIPS

An abstract is only as good as the abstracter. Whenever possible, confirm the information by looking at the original documents.

You can't count on published sources with unsubstantiated data—and this includes the Internet—to provide you with accurate information. You must always corroborate the claims made in these with other sources. Even reliable published sources can have errors. There are many steps to publication, and at any point mistakes may be introduced, particularly in dates.

Who supplied the information? It's unwise to depend solely on the information given to the census taker. You don't know who answered the questions. Was it the head of the household? A child? A neighbor? Maybe the census taker thought he knew the answers, so he didn't bother to ask. Maybe your grandmother shaved a few years off her age not wanting your grandfather to know she was older than he thought, or she didn't want her neighbor—the census taker—to know how old she really was. Human nature hasn't changed over the years; people make mistakes—both honest and deliberate.

The accuracy of the date on the tombstone depends on the knowledge of the informant and the skill of the stone cutter. In our example, the Bible record was made near the time of the event by someone who was an eyewitness or participant; it's probably the most accurate date out of those listed on the tombstone, census, and Bible record.

When the family seems to have had two or more children with the same name, consider the possibility that one child died and a subsequent child was given the same name (a common custom). Or that the family was German and used the same first name for all the boys, distinguishing them by giving them a different second—or middle—name.

Perhaps you've read an abbreviation as Jas. and assumed that was James, when actually the abbreviation was Jos. for Joseph. Or you couldn't figure out why the birth place for so many people was "do." When you learn that "do" was an abbreviation used for "ditto," you can go back to the record and find the actual birth place that was dittoed.

The Data Doesn't Conflict—There Isn't Any Data!

Your ancestor just seems to appear in town with his wife and children, and you have no idea from whence he came. Start looking at the records created by his neighbors. Families didn't move alone; they came with their friends and relatives. They came because of discord in their old church or because they wanted more land or new opportunities.

Imagine how difficult overland travel was through uncleared lands and over the mountains. Might they have come to the area by water from another area or through a natural mountain pass? Could they have crossed an ice-frozen lake in the winter to get to a new territory? Do their deeds tell you where they lived prior to now? Or perhaps county histories detail a migration of an ethnic or religious group from another county.

If you can't find a burial where you expect it, one of several things may have happened. Elderly parents may have moved closer to their children in another county. Widows and widowers, who moved to be near sons or daughters, may have died in the new county but were shipped back to the original home to be buried next to a spouse. Pioneers died en route to new homes and were often buried in unmarked graves along the way. In the twentieth century, cremations became more acceptable with remains scattered rather than buried.

Coaxing information about our female ancestors from the records can be especially frustrating. The general advice is to examine the records of the men in her life—her husband, brothers, brothers-in-law, and father. It's essential that you perform an exhaustive search through basic records available in print, on the Internet, and onsite.

TREE TIPS

If you can't find the maiden name of your great-grandfather's wife, look for families with female children in his immediate neighborhood to see if you can find some possible candidates. Look for more records of those families to see if they lead you to the right young woman.

Everything Is Relative

When the United States was primarily a rural society, the marriage pool for young men and women consisted of their neighbors within walking or horseback distance. Young men and women married the other young men and women with whom they grew up and saw at church and family gatherings. "Older men" of 25 waited for a beautiful 14-year-old to reach marriageable age. Because the marriage pool was so limited, first cousins married in areas where the law permitted. (This custom is no longer allowed.)

Families became intertwined when a brother and sister married a sister and brother from the same family, or when two sisters of one family married two brothers of another. These situations made their offspring "double cousins."

Many women died in childbirth, leaving a widower with several young children and perhaps a tiny infant to raise. He was eager to marry again, if only to have help with the children, and sometimes married a sister of his dead wife. Aunt Rachel became a stepmother to her nieces and nephews.

Happy New Year—March 25

When the glossy new calendars start arriving in December, it probably doesn't occur to you that New Year's Day was not always 1 January. Furthermore, it may not be obvious how a change in the calendar can affect genealogical research.

Calendars were developed to make sense of the natural cycle of time: days and years from the solar cycle, and months from the lunar cycle. It took some experimentation before folks got used to the current system. There are many calendars, but for right now, we need be concerned with only the Julian and Gregorian calendars.

The Julian calendar resulted from Julius Caesar's reformation of the system to conform more closely to the seasons. The Gregorian calendar was Pope Gregory XIII's solution for the gradual problem that had developed with the Julian calendar: over time, the calendar became 10 days off the natural solar cycle. To compensate for this, the Gregorian calendar dropped 10 days from October in 1582. To keep this problem of extra days from reoccurring, one day was added to February in every year divisible by 4 (called a "leap year").

The Gregorian calendar was adopted by different nations at different times. Generally, it was adopted by Britain and her colonies in 1752. The leap years meant the calendar was now 11 days out of sync with the solar cycle. To take care of this, the system was adjusted so the leap day is dropped from every century mark not divisible by 4. Instead of dropping 10 days in October, the British dropped 11 days in September and changed the New Year from 25 March to 1 January. Why do you need to be aware of this interesting bit of trivia? The calendar change makes dates prior to 1752 in the months of January, February, and up to 25 March subject to *double dating*.

DEFINITION

Double dating is the practice of giving dates that occurred before 1752, from 1 January through 25 March, two dates to represent the old and the new calendar; for example, 23 January would appear as 23 January 1749/50. Alternately, this may be shown as 23 January 1749 O.S. (old style) or 23 January 1750 N.S. (new style).

George Washington's birthday, 11 February 1731 under the old calendar, became 22 February 1732 under the new calendar. Therefore, his birth date would be expressed as 22 February 1731/32. This doesn't change his birth date; it changes only the way the date is expressed. In 1731, February was almost the end of the year because 1732 began on 25 March. After the year 1752, there are two things with which you must contend: the dropping of 11 days, and the change of the beginning of the year from 25 March to 1 January.

The calendar change affects your research because it's sometimes hard to determine whether the dates were recorded in the old or new style. You may think the change wasn't significant enough to make a difference in your research, but it likely will. If you find records that indicate Abraham was born on 27 March 1741 and his younger sister Ruth was born on 23 March 1741 you may think there's something wrong. In reality, the dates are likely correct, because 23 March 1741 in the old style calendar followed 27 March 1741 by about 12 months.

The change in calendar can also explain the seemingly erroneous court item that shows a will dated 3 December 1740, and proved in court 1 January 1740. "That can't be," you think. But it can and was, because 1 January 1740, followed 3 December 1740 (expressed now as 1 January 1740/41). Once you understand this, you need to show it in your records, or others will think you've erred. This is the best way to show it: "The will was dated 3 December 1740, proved in court 1 January 1740/41." Be sure to use a slash and not a dash; it's the slash that clarifies you're referring to the double dates caused by calendar changes.

Don't just convert the date with no explanation. If you do prefer to express it in new style (N.S.), show it as "1 January 1741 N.S." so others understand under which calendar you've given the date. For more on calendar history see infoplease.com/ipa/A0002061.html#A0880641.

TREE TIPS

Though the difference of 11 days can explain some records (he died on the third of the month but wasn't buried until the thirteenth), genealogists shouldn't convert dates to account for the 11-day difference unless the old-style date would cause confusion. If it does, change it and indicate that you've done so to conform to the new-style calendar, or leave it as is and explain the seeming discrepancy.

It's important to interpret carefully when faced with dates that were shown in months. This prevalent Quaker custom was also used by some others. A date of "30 9ber 1741" (or "30 9 mo 1741") is 30 November 1741, based on the calendar of the time in which the first month was March and the ninth month was November. It's best, in extracting records that are expressed in months, to write them in your abstract as shown in the original record. If you want to show it also as it would be under the present calendar, add that in brackets. For example, John Betts was born, according to the Quaker Monthly Meeting record, 30 9ber 1741 [30 November 1741 N.S.].

Rushing to Conclusions

Other things besides conflicting or missing data and new dating systems can throw off your research. Pay attention to the clues. If you find William and Mary in the 1860 census record with three children, and in the 1870 census record you find William and Polly with five children, don't assume Mary and Polly are one and the same, even though you know Polly is a nickname for Mary. They may indeed be the same woman, but analyze the record. Is the woman on the 1870 census approximately 10 years older than on the 1860 census? Are the birth places the same for both names? Are the children's ages consistent with one marriage, or is there a gap between numbers three and four, or four and five? If the older children are teenagers and the younger is three years old, there may have been two marriages.

The will you find for James makes bequests to "my wife" and "my children," all named. Don't assume the named wife is the mother of these children. She may be the mother of all of them, some of them, or none of them, but this wording doesn't tell you.

How about the census record? You found your ancestor John Matson listed as age 30, and with him is Mary, age 28, and children George, age 5, Martha, age 3, and Sally, age 1. You write in your records, "John Matson was born ca. 1820, married Mary—born ca. 1822, and had children George, born ca. 1845; Martha, born ca. 1847; and Sally, born ca. 1849." What's wrong? You've just assumed Mary was his wife! It's likely she was his wife, but the 1850 census record doesn't include relationship. If further examination of the family records proves perplexing, consider that his wife may have died. Perhaps a relative with the same surname came to help out with the small children. Keep all possibilities in mind if the census record doesn't match other known records of the family.

TREE TIPS

Does finding a record with the title "Colonel" send you searching immediately for the military record of his appointment? Pause and consider; is this colonel a southern plantation owner or a man of some substance in the town? If so, "Colonel" may be an honorary title of respect. He may have had some military or militia experience, but may not have attained the military rank of colonel.

You find two men with the same name, except one is designated Senior, and one Junior. Don't assume they're father and son. Junior and Senior may have been used to distinguish individuals in the community with the same name. They may be father and son, but it's just as possible they're cousins, uncle and nephew, or totally unrelated. The only thing you know for sure is one is older and one younger. Additionally, when "Senior" dies, "Junior" may be promoted to "Senior," further confusing us as we try to determine the correct line. If a man with the same name came into the area who was older than the original inhabitant, the town may designate the original resident as John Smith I and call the new resident John Smith II.

You discover a will in which John refers to his "now wife." It could mean he was married previously. But it usually means nothing more than designating his wife at the time he drew up the will, rather than a woman he might marry after the will was signed if he were to divorce or become a widower.

Theories of Relativity

Sooner or later—usually sooner—you're going to find records that suggest there are two people with the same name in the same place, only one of whom is your ancestor. How to sort them out?

One method is to write each thing you've found on or in a spreadsheet. Watch for connecting events, or inconsistencies and contradictions. Remember that you're researching not a name, but a person. Look for associates, occupations, church affiliations, or anything that will develop identities for the people you found who happen to have the same name. Look at their neighbors, too.

Deed: In 1822, Thomas Dunn sold 100 acres in Decatur County, Indiana. Land adjoins that of Robert Johnson and William Ballard.

Census: In 1850 census record in Decatur County, Indiana, Thomas Dunn, farmer, age 30, wife Elizabeth, children James and Elizabeth.

Military: Thomas Dunn served in the War of 1812.

Will in probate file: Dated 20 July 1852, and probated 1852, Thomas Dunn's will bequests to wife Sarah, son James, daughter Elizabeth; witnesses to will were Robert Johnson, William Ballard. Admitted to probate in December 1853.

Begin to build a mini-biography for these men. The Thomas Dunn who sold land in 1822 can't be the Thomas Dunn age 30 in the 1850 census record because he would have been two years old in 1822. (To sell anything at age two, a guardian would've done so on behalf of the individual.) The Thomas Dunn age 30 on the 1850 census record can't be the Thomas Dunn who served in the War of 1812 because he wasn't born yet, but the Thomas Dunn who sold land may have been the same one who served in the War of 1812. The Thomas Dunn whose will was probated in 1852 may be the same man who sold land because the witnesses to his will owned land adjoining the land he sold in 1822. Who are the neighbors of the Thomas Dunn listed in the 1850 census record? Are there any Johnsons or Ballards?

Your problem may be more difficult than this, but the approach is the same. Sort out what you have, and make the connections as best you can.

Based on the Evidence ...

In the preceding example, the *evidence* is the deed, the military record, the census record, and the will. They're all useful in proving or disproving there were two Thomas Dunns.

DEFINITION

Evidence is the information offered as proof of a claim or answer to a question relating to a lineage or relationship. All else is only information or bits of data that may or may not later be proven by evidence.

You will often be faced with three or four records of a specific event, all differing. The Bible shows Hattie was born 3 February 1851. Her tombstone shows her birth date as 3 February 1852. Her death certificate, filed in 1915, gives her birth date as 3 February 1852. Which is correct? Do you decide that because two records showed 3 February 1852, that's the correct date?

Consider three questions: *WHO* gave the information; *WHY* did they give it; and *WHEN* did they give it? The answers will assist in determining which is the most likely date.

It's not unusual to see the death record, obituary, and tombstone all showing the same date, even if the date is incorrect. It's likely the same person supplied the information for all three. If that person had the date wrong, all three records will be incorrect. Fall back on *WHO*. Was it someone who was in a position to have the correct information? Or was it a daughter who never knew her mother fudged her birth date all those years?

Consider the Bible. *WHEN* was the Bible published? If the title page on the Bible was 1848, and the family entries start with the marriage of the parents in 1849, followed in the same handwriting by the births of the children, it seems likely the couple acquired the Bible when they were married and subsequently added births as the children were born. Because that was done close to the time of the event (answering the *WHEN* question) and by someone in a position to have the correct information, it would appear to be correct.

We then ask *WHY*. In this case, there doesn't appear to be any reason why the Bible date might be purposely entered incorrectly. (On the contrary, if a record was being provided to establish a pension or Social Security, and didn't jibe with other records, you might consider the *WHY* important.)

The Genealogical Proof Standard to the Rescue

Developed by the Board for Certification of Genealogists, the Genealogical Proof Standard has become the credibility standard for the genealogical community. As explained in *The BCG Genealogical Standards Manual*, the standard consists of a five-step process:

1. We conduct a reasonably exhaustive search for all the information that is or may be pertinent to the identity, relationship, event, or situation in question.

2. We collect and include in our compilation a complete, accurate citation to the source or sources of each item of information we use.

3. We analyze and correlate the collected information to assess its quality as evidence.

4. We resolve any conflicts caused by items of evidence that contradict each other or are contrary to a proposed (hypothetical) solution to the question.

5. We arrive at a soundly reasoned, coherently written conclusion.

Apply those five to your genealogical research as you seek to prove your ancestral relationships.

Proof Summaries and Proof Arguments

We include in our compilations proof arguments and proof summaries so others can see how we arrived at our conclusions. For guidance in writing those arguments or summaries, see the excellent article "Skillbuilding: It's Not That Hard to Write Proof Arguments," by Barbara Vines Little for the Board for Certification's September 2009 issue of *OnBoard*. The article is available online at bcgcertification.org/skillbuilding/skbld099.html, and it can be printed for referral as you work to prove the relationships in your genealogy.

Reaching a Sound Conclusion

There are basically three ways to build a case in genealogy. You can construct it with pieces of direct evidence, pieces of indirect evidence, or a combination of both. You measure the reliability of what you've found by asking: Are the sources original or derivative? Is the information primary or secondary? Is the evidence direct or indirect?

Elizabeth Shown Mills' *Research Process Map* is a clear guide to evidence analysis and will help you evaluate the quality of what you've found. The map is available as a laminated QuickSheet from the Board for Certification of Genealogists and you can order it online at bcgcertification.org/catalog/index.html. Also available at that website is the concise guide, *Genealogical Proof Standard: Building a Solid Case* (the 3rd edition) by Christine Rose. Explanations and examples are given of both the Genealogical Proof Standard and the evaluation of evidence to support that standard.

The message of this chapter is "don't despair." You'll constantly find discrepancies in information, and will frequently lament the absence of documents providing the precise answer. There *are* ways to overcome these challenges!

The Least You Need to Know

- When the data conflicts, weigh the evidence.
- Write mini-biographies to sort out individuals with the same name.
- Don't read into the records more than is there.
- Use the Genealogical Proof Standard to build a solid case.
- There are concrete ways to evaluate evidence.

Expanding Your Horizon

Now you're really hooked, and starting to look at all the things you'd like to have: books, magazines, and equipment. There are even national conferences, regional seminars, institutes, and courses. Where to spend your money?

You'll also learn a bit about sources that won't be your first line of attack, but that you'll need as you dig deeper. Records on orphans, your Native American ancestors, the naturalization process—these and much more await your discovery. And we'll introduce you to the newest tests now available for DNA samples.

Finally, with some strategies for success and some words of caution, you are well on the road to many happy years of climbing your family tree!

Spending Your Money Wisely

In This Chapter

- Analyzing your needs
- Allocating your expenditures
- Acquiring references gradually

Your genealogical education is a lifelong process. Everything you learn leads to areas where you need to delve more deeply. Much of what you want to know is available without cost through public libraries, but there are also many educational opportunities for which there is a charge. You start wondering how you can afford to acquire all the knowledge and resources you need. Don't despair; there are many valuable ways to use your money.

Just as you apportion your money for the necessities of life, budget for your interests. Consider these five areas:

- Education, classes, and basic reference materials
- Memberships and subscriptions, including those online
- Services
- Resources including books, computer software, and Internet access
- Luxuries such as the books you constantly consult at the library, and CD-ROMs that save you time

Readin', Writin', and 'Rithmetic

Seek every opportunity to attend classes and lectures on genealogy. Become familiar with the names of outstanding genealogists, look for opportunities to hear them lecture, and read their published articles.

In addition to the seminars sponsored by your local genealogical society, there are regional and national seminars and conferences. Attending the conferences will give you new ideas for solving your problems, help you hone your skills, introduce you to new materials, and connect you with others who share your enthusiasm for genealogy.

> **TREE TIPS**
>
> Regional conferences aren't on a regular schedule, so watch for announcements about those to be held in your region. The dates and locations are published in genealogical magazines and are sometimes posted on library bulletin boards. You can also find them in the online calendar at the Federation of Genealogical Societies (FGS) website, fgs.org/calendar/index.php. See upcoming events at Dick Eastman's egon.com by scrolling down the right side and clicking on "Upcoming Events." There are other events sites listed at cyndislist.com; click on "Categories," and then select "Events & Activities" from the list of options. The Genealogical Speakers Guild (genealogicalspeakersguild.org) will also have listings.

The National Genealogical Society (NGS) at ngsgenealogy.org and the Federation of Genealogical Societies (FGS) at fgs.org each sponsor a national conference in a different location every year. These conferences run for three or four days, and feature six or seven simultaneous presentations by top genealogists and lecturers every hour. The topics vary widely: military files, land records, problem solving, ethnic research, newspapers, courthouse research, and methodology, to name a few. Presentations on DNA, digital cameras, and scanners integrate technological advances with genealogy.

Your conference registration includes a syllabus, either a thick volume or a CD-ROM, with material submitted by the lecturers. The materials range from a few paragraphs to a detailed outline of the lecture, including bibliographies, maps, special instructions, and more.

> **TREE TIPS**
>
> The syllabi are good references for future research. As you tackle a new problem, see if you have information from any lectures pertaining to it, and study the bibliographies. They may give you ideas on information sources. Even if you didn't attend a conference, you may find the syllabus helpful.

Networking

Networking is a vital part of the national conferences. In talking to your tablemates at lunch or relaxing during a break, you may find you share a common research problem or possibly even an ancestor. Many attendees have made connections from such contacts.

Exhibits

National conferences draw hundreds of attendees. In addition to the lectures, one of the attractions is the exhibit area where vendors sell everything from rare books to T-shirts with genealogical messages. You'll find genealogical supplies such as charts alongside materials for preservation such as acid-free paper. Books (new and used) and maps are big sellers. Some vendors specialize in photo restoration. Large genealogical societies sponsor booths, recruiting members and selling publications. National and local research libraries and archives are well represented. The biggest presence seems to be from companies offering online subscriptions. Government agencies such as the National Archives and the Bureau of Land Management often have booths, too.

Hands-On

At both the regional and national conferences, there are usually some hands-on workshops that enable you to develop and practice skills important to your research. Practical experience under the guidance of knowledgeable, patient instructors makes the techniques easier to acquire.

Occasionally, there are workshops devoted to abstracting and handwriting, as well as workshops on how to complete successful lineage society applications. In the latter, the emphasis is on the types of evidence needed to establish your line, and suggested sources for finding that evidence.

Computer workshops concentrate on specific genealogy programs, explaining many of their features, or on general topics, such as learning to use the Internet for genealogical research.

Institutes

For a week of saturation in genealogy, attend one of the various genealogical institutes. Rather than a series of unrelated lectures, the institutes usually offer one or more tracks of related subjects. For instance, at the Institute of Genealogy and

Historical Research at Samford University, Birmingham, Alabama (samford.edu/schools/ighr), there may be tracks for introductory genealogy, intermediate and advanced methodology, records of the South, military records, genealogical writing, and others. Some tracks are offered yearly; specialty tracks are available periodically.

The National Institute of Genealogical Research (NIGR), Washington, D.C. (rootsweb.ancestry.com/~natgenin), takes place at the National Archives and concentrates on genealogical materials available in the Archives. The Utah Genealogical Association sponsors the annual Salt Lake Institute of Genealogy (infou.org). Some genealogists return to these institutes year after year, taking advantage of the variety of courses offered each time.

> **TREE TIPS**
>
> Brigham Young University has a four-year college curriculum culminating in a bachelor's degree in Family History. This requires class attendance on campus at Provo, Utah. Details are at universities.com/edu/Bachelor_degree_in_General_Studies_Family_History_at_Brigham_Young_University.html.

Online and By Mail

The National Genealogical Society (ngsgenealogy.org) offers several online courses. They also offer an accredited and highly rated Home Study Course. Completing the lessons gives you a good grounding in how to find and record your sources, maintain your records, and evaluate your evidence.

For online seminars offered by the New England Historic Genealogical Society, go to americanancestors.org/online-seminars. A number of diverse topics are included, such as *Civil War Pension Research: Union Soldiers*, *Researching Your Newfoundland Ancestors Part One*, and *Methods of Finding a Wife's Maiden Name*. These and many others are listed. There's no charge to view this series on their website.

For more online opportunities go to cyndislist.com, click on "Categories," and then select "Education (Genealogical)" from the listing. Improve your skills from the comfort of your home.

Being a Joiner

Allocate part of your genealogy budget to society memberships and subscriptions. You'll usually benefit from belonging to at least three organizations: one local, one state or regional, and one national. As you become aware of more areas where your ancestors lived, consider joining societies there. The newsletters and journals you receive as part of your membership dues are a part of the continuing education process that's so essential to successful genealogy. To locate societies, check the FGS's Society Hall Directory at fgs.org/societyhall; you're sure to find several of interest to you among the hundreds listed.

Consider subscribing to some of the premier genealogical journals that publish the results of scholarly work. Even if the information presented has nothing to do with the surnames you're researching, the methodology used is important and may eventually help you solve a problem. Genealogists often develop an interest in obscure records or subjects; the articles they write about those interests may lead you to new sources. Journal articles can alert you to newly published abstracts or compiled genealogies. Among these scholarly journals are *The Genealogist*, the *New England Historical and Genealogical Register*, the *New York Genealogical and Biographical Society Record*, the *National Genealogical Society Quarterly* (*NGSQ*), and the *American Genealogist* (*TAG*). Enter the society's name in your browser to locate their websites.

An important part of journals is the book-review section with its evaluations of recent publications. The books may pertain to a specific region, a family, general reference, related topics, or other areas of interest. Reading the reviews may suggest books that would be useful in your research—and those to avoid.

In addition to the scholarly journals, there are popular publications with articles of general interest. They often include question-and-answer columns on different topics, such as ethnic research, computers, or a particular geographic location. Other periodicals focus on one subject, such as computers or Irish genealogy.

TREE TIPS

Don't overlook the footnotes or bibliographies accompanying journal articles. They're rich sources of new ideas for your research.

Are You Being Served?

Genealogists use many fee-based services that are priced reasonably, but that add up to substantial amounts over time. Budget yearly for photocopying, postage, and loans of books or films through interlibrary loan or rental companies. If you choose to do online research but do not already have Internet service in your home, you will need to budget for that as well.

Occasionally, you may need professional help with a stubborn problem. The *Directory of Professional Genealogists*, published by the Association of Professional Genealogists, is online at apgen.org; the *Certification Roster*, published by the Board for Certification of Genealogists, is at bcgcertification.org. Lists of accredited genealogists for specific geographic areas can be obtained at icapgen.org, the site of the International Commission for the Accreditation of Professional Genealogists.

It may be more economical and efficient to hire a researcher than to travel to a distant area where you're unfamiliar with the resources. In some counties, the officials are too busy with the daily work of the county to engage in research, so they provide a list of researchers. Many archives and libraries also maintain researcher rosters. These lists are usually for the convenience of inquirers and aren't meant as endorsements of the researchers' skills.

Before hiring genealogical help, know what to expect. The Association of Professional Genealogists developed a brochure, *Why Hire a Professional Genealogist?* It has tips on how to find a professional, how to evaluate the credentials, and what to expect in the way of costs and results. Download it from apgen.org/articles/hiring_a_professional.pdf.

Developing Your Home Support System

Add to the references and resources you keep handy at home. Tapes, books, catalogs, computer programs, Internet subscriptions, and CD-ROMs can advance your research from the comfort of home.

Most national conference lectures, and some regional ones, have in the past few years made most of their lectures available on CDs, which you can purchase. Although you miss the visual parts of the lectures, the CDs are nonetheless useful aids for your education. Peruse the online catalog at jamb-inc.com.

Adding to Your Library

Build your collection little by little. Start with some basic guidebooks, and add to your collection as you progress in your research. You may find there's one book you regularly consult at the library; consider investing in a copy for your own bookshelves. When you find a geographic area where you're doing a great deal of research, you'll probably want to purchase books about that area—not just genealogical books, but also histories. If you find you have an interest in a specific ethnic group, look for books in that field. There are numerous books on many aspects of Jewish genealogy, as well as on African American, Italian, German, Russian, Scottish, and so on. To learn about an ethnicity or cultural group, attend lectures or study the bibliographies in conference syllabi and journal articles.

> **TREE TIPS**
>
> Don't concentrate only on genealogical offerings. Social histories that deal with particular geographical areas or time periods can be quite useful, especially when you start writing narratives about your ancestors. You may learn stories about your ancestors while reading the history of an area.

Bookstores

There are several large booksellers that specialize in genealogical materials. Their owners and managers are genealogists who understand the needs of the field. Some specialize in books that aren't generally thought of as genealogy books, but that can aid in your understanding of the laws, customs, and conditions that may have influenced your ancestors' behavior.

Don't overlook the major societies and archives. Many of them publish historical books and finding aids for their collections. North Carolina Archives, the National Archives, the New England Historic Genealogical Society, and the National Genealogical Society are among them.

Booksellers and many societies and archives advertise in genealogical publications. Look for them and send for their catalogs or reach them through their websites. For online bookstores, search services such as AbeBooks.com or eBay.com can be good sources for rare or special-interest books or CDs. AbeBooks, an umbrella organization for thousands of book dealers, can be searched either in a simple search or an advanced search; try both. They also have a useful "want" list so you can enter titles you are seeking even if they don't presently have those titles.

Feeding Your Computer Some Mapping

Mapping programs are useful for converting a legal description of land to a graphic representation. DeedMapper (see information at directlinesoftware.com) enables you to plot the land. See also goldbug.com (mentioned in Chapter 13).

Links to numerous type of maps can be found at lib.utexas.edu/maps/map_sites/ map_sites.html. Another site with thousands of images is The David Rumsey Map Collection (davidrumsey.com), a useful site you will return to often.

Google Maps (maps.google.com) can quickly put the address you seek onto a map. For a variety of city and town maps, military parks, and many others, go the Library of Congress website at loc.gov/rr/geogmap/guides.html.

Using Your Computer to the Utmost

When inspiration strikes in the middle of the night, your subscriptions to online sites permit you to confirm, refute, or enhance your ideas on the spot. Definitely allot some of your genealogy funds to subscriptions including census records. When deciding on a subscription service for access to census records, weigh the value to you of the services of other collections such as historical newspapers, British records, French-Canadian records, and others of that ilk.

Before Internet research became commonplace, a lot of information was published on CDs. They take advantage of one of the things computers do best—quickly searching through mountains of information. As with any publication, however, be wary. Many CD compilations are incomplete or have errors, just as many printed indexes do, and many of them are out of date. Still useful, though, are CDs with fully digitized, every-word searchable, scholarly journals from their inception up to the present as well as long-out-of-print books. For example, you don't need to even get to a library to look up an article in the New York Genealogical Society's *Record* from 1870 to 1960. The images are available on five CDs. Just visit their website at loc.gov/rr/ geogmap/guides.html to purchase one or all of the discs. Be sure before purchasing CDs that they will work with your computer's operating system.

Budgeting for Genealogical Luxuries

All these resources can move into the luxury category as genealogy takes hold of your life. You find you want to attend all the conferences and institutes and load up on memberships, subscriptions, books, tapes, and computer programs. You'll covet the latest multivolume indexes available on CD. You'll buy into the reasoning that a copier and a microfilm reader will save you time and money.

If you don't have a computer, you'll want one. If you have a computer, you'll want a scanner, a digital camera, a smart phone, and more … and more. If you have a laptop, you'll hear the siren song of travel accessories: a tiny mouse, a travel keyboard, a miniaturized surge protector, security cables—the list lengthens.

Surely you'll be tempted to take a genealogy cruise. Imagine enjoying a vacation in the Caribbean or Alaska or Mexico with all that implies, as well as attending onboard lectures presented by prominent genealogists. Exchanging ideas with other attendees in a social setting can supercharge your research, but the price can be somewhat steep.

Before you go on a spending spree, study your budget. Be an informed consumer, first buying the essentials that will help you become a skilled researcher. Ask yourself, "Will this purchase add to my genealogical skills and knowledge?" "Will it help me trace my ancestors?" Later, you can contemplate the purchase of the big-ticket items.

The Least You Need to Know

- Budget for your genealogical necessities and luxuries.
- Make taking classes and acquiring books a priority.
- Always ask yourself if a purchase will contribute to your genealogical knowledge.

DNA—Why the Hype?

In This Chapter

- How to use DNA tests for genealogical purposes
- How DNA projects can connect you to relatives
- How to order and take a DNA test
- How DNA testing and genealogy came together

Do you sometimes wish you could look into a crystal ball to see who your ancestors were? Do you ever regret not asking more questions about your relatives when your parents, grandparents, or great uncle were alive? Do you spend time daydreaming, or tossing and turning at night because you don't know the identities of your birth parents? Are you interested in meeting distant cousins and other living relatives?

If you've answered yes to any of these questions or have run into brick walls with your genealogy research, consider a deoxyribonucleic acid (DNA) test. Nearly every cell in your body contains DNA, and they carry the genes of your ancestors. While crystal balls may not be a practical solution, DNA testing is.

Anyone can take a DNA test, and you don't have to be a scientist to understand your test results. Before ordering a testing kit, however, it's important to determine which DNA test will provide the answers you seek.

Genetics and Genealogy (G&G)

Using DNA testing for genealogical purposes has become a popular and common practice. It provides a means of connecting with others who share your same DNA. It can determine whether two people share a common ancestor, and provide a level of scientific certainty of the genetic relationship between two or more individuals. The results from a DNA test can validate your genealogy research or point you in the right direction.

For many years, DNA has been used to identify people in criminal, forensic, and paternity cases, but tests for genealogical purposes were not available commercially until 2000. Advances have occurred at a very rapid pace since then, though limitations still exist.

There are three main types of DNA tests available for genetic genealogy research: the Y Chromosome (Y-DNA) test, the mitochondrial DNA (mtDNA) test, and the autosomal DNA test. Each of these tests is very different and provides you with a unique set of information. If you're a female, it is important to note that gender also plays a role in which kind of test you can take.

Following is a list of some basics you'll need to know before ordering your testing kit:

- What kinds of tests are available?
- What kind of information will each test provide?
- Who can take each test?
- What is in a testing kit?
- How do I order a test?
- Can I get a discount?

DNA Tests Available for G&G Research

Although all ancestors pass on autosomal DNA to their descendants, due to recombination (meaning its pattern of inheritance), autosomal DNA contains information about some but not all of one's ancestors.

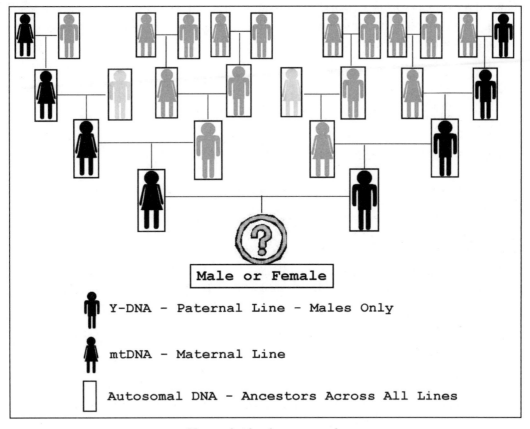

How to decide what test to order.

Y-DNA Test

The Y-DNA test looks at markers on the Y chromosome, and traces the paternal line (a male, his father, his father's father, and so on). Only males have Y-DNA, which they inherited from their fathers and will pass on to their sons. If you're a female you cannot take this test, but you can order a test and ask a male relative who descends from your paternal line to take a test for you (a father, brother, uncle, nephew, or cousin). The results of your male proxy will provide you with all the information concerning your paternal ancestry that you would have obtained if you had been the male who submitted his DNA. The Y-DNA test is *the* test that made DNA testing popular for genetic genealogy research. Since Y-DNA is passed down virtually unchanged from father to son for many generations, a genealogist can use the results from a male's Y-DNA test to trace a paternal lineage back hundreds of years into the past, and sometimes further.

mtDNA Test

The Mitochondrial DNA (mtDNA) test examines DNA in *mitochondria*, and traces the maternal line (your mother, her mother, and so on). Both males and females can take this test to trace their maternal line.

> **DEFINITION**
>
> **Mitochondria** are structures found in every cell in your body, which contain their own DNA and essentially serve as the power pack of the cell.

Three regions of the mitochondria are used for mtDNA testing: Hypervariable Region-1 (HVR-1), Hypervariable Region-2 (HVR-2), and the coding region. Although maternal lineage is usually more difficult to trace because females typically change their names upon marriage, diligent genealogical research can usually identify distant females in your ancestry.

The mtDNA test is a very useful tool when the paper trail for your female ancestor stops, but you can trace the ancestry of a female you believe to be your ancestor's sister. Comparing your mtDNA results with a descendent of that suspected sister can prove or disprove whether the two women did indeed share a common maternal ancestor.

Autosomal Test

The autosomal test looks at autosomes, which are markers on 22 pairs of chromosomes that aren't linked to gender. Autosomal DNA makes up the vast majority of your DNA, and is inherited from each parent equally. Both males and females can take this test to look for DNA matches with relatives from all of their lines.

The autosomal DNA test is referred to as the "Family Finder Test" at Family Tree DNA (familytreedna.com), and the "Relative Finder Test" at 23andMe (23andme. com). This test is rapidly gaining popularity with genealogy researchers for its ability to cross the gender barriers, and its usefulness in discovering relationships with others in more recent generations. Autosomal DNA from each parent uniquely recombines with the conception of each child (with the exception of identical twins).

As a result, your autosomal DNA is a random recombination of selected autosomal DNA of your parents, which includes random recombination of autosomal DNA from their parents, and so on. Due to its unique inheritance pattern, autosomal DNA contains genetic information from some but not all of your ancestors.

This test can predict the degree of relationship between two persons and identify a genetic connection between two people who would otherwise have no documented proof of a relationship.

TREE TIPS

While an autosomal test is not intended for use as a replacement for traditional genealogy research, if you use it in conjunction with other DNA tests or a documented paper trail, it may lead you to discovering relatives you've never met, and to finding new information about your more recent ancestors.

Choose Wisely

Choose a testing company that offers a wide variety of tests and free storage for your DNA samples. Like many others, you may want later to order additional tests or upgrade the one you took. All of these tests, when used in combination with one another, can potentially take you further in your research than one type of test alone. An mtDNA test can help you determine whether one of your matches from your autosomal DNA test is related to you via your direct maternal line. Similarly, if you match with the same person on your Y-DNA and your autosomal DNA tests, regardless of their surname, you can be assured you're in fact related through your paternal line.

DNA Testing Companies, Kits, and Prices

As the main purpose in taking a test is to compare your results with others, the success of your test can be dependant on the number of people in a testing company's database. For this reason, I recommend you purchase a DNA test from one of the large DNA testing companies. Some of the large testing companies include the following:

- Family Tree DNA: familytreedna.com
- DNA Ancestry: http://dna.ancestry.com
- 23andme: 23andme.com
- Oxford Ancestors: oxfordancestors.com

The large DNA testing companies have extensive websites that offer tutorials and other aids to assist their customers in learning more about the types of tests they offer.

Save Your Pennies

The cost of a DNA test varies between DNA testing companies and the type of test(s) the customer orders. A minimum of 37 to 46 markers is usually required for a meaningful interpretation and comparison of a Y-DNA test; anything less could result in a large number of false positive matches with individuals who share similar DNA but who are not related within the last several thousand years. Generally speaking, the more markers you order, the more accurately you can determine your relationship to those with whom you match.

TREE TIPS

Consider ordering or later upgrading to a 67-marker Y Chromosome test, which can further define those who are closely related to you. Adding the mitochondrial HVR-2 test to an HVR-1 test is also a good investment in narrowing the number of persons who share your common maternal ancestor. Today, many are recommending a Full Mitochondrial Sequence (FMS) test.

Because most of the DNA testing companies test different markers for their Y-DNA tests, it's usually difficult to compare the results of individuals who tested with different companies. Family Tree DNA offers a 111-marker test that makes it possible to compare your test with others who have tested elsewhere.

The more markers you order for the Y-DNA test or regions you have tested for the mtDNA test, the more information you have to compare to other individuals or groups within a surname, mtDNA, or geographical project. The amount of information you receive from your test results determines how accurately your relationship to others the test can determine, and can therefore determine the success of your test in discovering the answers for which you're looking.

Can I Get a Discount?

If you join a project at Family Tree DNA, regardless of the type of test you order you may qualify for a group discount rate on your DNA testing kit.

DNA testing companies generally offer discounts to new customers on combination kits, as well as to existing customers who order additional tests and upgrades on their kits.

Family Tree DNA offers occasional promotions on its Facebook page at facebook. com/FamilyTreeDNA, which provide substantial discounts on testing kits. If you have a Facebook account, and would like to receive notifications of special offers, click "Like" on their Facebook page. You don't have to be a member of Facebook to take advantage of these sales. Simply keep an eye on this company's Facebook page, and watch for an announcement and a coupon code.

Taking the Plunge

Testing is best done through individual surname projects whenever possible. Surname projects are genetic genealogy projects that assist members in tracing their paternal lines. Generally, members in these projects share the same or similar surname. Because Y-DNA is inherited from father to son, like a surname, it is ideally suited for surname studies. Since late 2000, thousands of surname projects have been established by interested groups to create opportunities for individuals to collaborate and explore their common ancestry.

Because Y-DNA test results are most useful when one can compare the results to others with the same or similar surname, it is especially beneficial for you to order a test from a company that hosts a DNA project with your surname. The testing company may be able to advise you, once your test results are complete, as to whether an mtDNA lineage, *geographical project*, or autosomal DNA projects are available for you to join.

DEFINITION

A **geographical project** connects individuals who live in a specific area, or whose ancestors came from a specific country or region.

Administrators of surname projects have access to DNA test results conducted on the surname you're researching, as well as tools to compare their members' test results. They usually have close ties with family associations and expert family researchers. It's important to note this because DNA results only become meaningful when coupled with traditional genealogical research.

Generally, each surname project is operated by an administrator using a single DNA testing company. To save money when ordering your kit, and to receive the most from your test results, order a kit through one of these projects.

If a DNA project covering the surname you're researching already exists, I strongly recommend you take part in it. If you proceed solely on your own, there are no guarantees that you will have access to other members' test results, or information about their lineage.

Lists of DNA projects can be found on the Internet at the following websites:

- Family Tree DNA: familytreedna.com/projects.aspx
- World Families: worldfamilies.net/search
- Cyndi's List: cyndislist.com/surnames/dna

TREE TIPS

If there is no project available for your surname, try searching for and joining a project with your mother's or a grandmother's maiden name. Many administrators of surname projects will allow family members with other surnames to join their projects, to qualify for the discounted rates.

Want to Start a Project?

If you checked the websites of the major DNA testing firms and didn't find a surname project, start one! The testing company may require a minimum number to set up the project and offer their discounted pricing, so check on that at their website. If you're interested in becoming the Surname Project Administrator, read all you can on the subject. There is considerable material on the web at the websites mentioned in this chapter. Check other websites, too, by entering the term "Y-DNA" in your browser and following some of the links that come up. See also the excellent books by Megan Smolenyak and Ann Turner, *Trace Your Roots with DNA*, and Thomas M. Shawker's *Unlocking Your Genetic History: A Step-by-Step Guide to Discovering Your Family's Medical and Genetic Heritage*.

Try starting a project with a particular goal in mind. Have a male line proven back to a particular progenitor that's well documented. Test an eligible male descendant from that line. Then test males who aren't sure of their ancestry but think they might be of that family, and see if any of them match. If DNA matches the uncertain descendant with the certain descendant, not only will the descendant be excited to identify his ancestral lineage, but it will also generate more interest in the project. Broaden the testing to include more persons, and families with similar surname spellings.

TREE TIPS

Most project administrators keep track of all the results. Some make personal contact with those tested, and discuss the results of the test. If this is too much for one person, have two people work on it. Family Tree DNA has a free and confidential online "public page" available to each project manager, where results can be posted and, if desired, identified by the name of the earliest known ancestor.

How Do I Order a Kit?

Once you've determined which kind of test you'd like to take, you need to order a testing kit from a DNA testing company. It is the company, rather than the type of test you order, that will determine the kind of testing kit you receive.

Order a testing kit by going to the website of the company providing the surname project of your choice and contact the surname project's administrator. Provide your address, specify which test you would like to order, and note how you prefer to pay for your kit. The administrator can send the kit either to you or directly to the person who'll be providing the sample for you.

TREE TIPS

Generally in surname projects, the results of a Y-DNA test are sent to the person who took the test, the person who paid for the test, and the project administrator.

Also provide the administrator the identity of the earliest proven ancestor in your line, and some basic information about him such as dates and locations. (This will assist the administrator in grouping your results with others.)

What's in the Kit?

Within a few business days of placing your order, depending on which company you use and the type of test you ordered, you'll receive either a cheek swab or a saliva collection kit in the mail. With it will be a complete set of instructions, a waiver you must sign and return with the completed kit, and a special return addressed envelope for the samples. Always keep in mind that although there are high percentages of meaningful matches, there's no guarantee your test will produce one.

Note: a meaningful match isn't guaranteed for any test. If your relatives have not yet taken part in a DNA test, then you won't have any matches. As time goes along and DNA testing expands, however, you'll eventually receive some matches.

On Pins and Needles

If you are like most, you'll be excited and impatient awaiting the results. Will you finally know to which family you belong? Will you learn about your ethnic background? Will you immediately find relatives?

It generally takes four to six weeks after you send your kit to the lab to receive your test results. Most companies allow you to sign into an online password-protected account to review your results. Some companies also send a copy of your results in the mail.

TREE TIPS

If you provide DNA testing companies with an email address, they will send an email notification to let you know when they have received your kit. They will immediately let you know via email when your test results are complete, and post your results on your personal page. Most of the testing companies also give you the names and email addresses of those who match you, provided you and the others have signed the consent form. (The company 23andMe is an exception because of privacy issues involved with medical information, which is this test's main focus.)

The project administrator also receives a copy of your results. If you ordered a Y-DNA test, the administrator compares the results with all others of the same surname and then reports to you any matches or similarities they find, together with relationships of earliest known ancestors.

What Is DNA?

If you don't want the technical stuff, skip these paragraphs! Most of you, however, will want to know a bit of what DNA's all about.

DNA is short for deoxyribonucleic acid. It carries the genetic code for all organisms, and provides the capacity for recreating them from single cells, to and through the procreation process.

In essence, the egg is a cell with a nucleus containing only half of 23 chromosomal pairs. In conception, the sperm penetrates into the nucleus of the cell, "fertilizing" it, or supplying the other half of each of these pairs. These pairs all combine by spiraling around each other, and exchanging small portions throughout.

The strands of DNA making up the chromosomes are in the general shape of a spiraled double helix, which looks like a twisted ladder. The fertilized egg then starts reproducing itself, ultimately forming the body with all its integral parts and organs.

The human body is made up of trillions of cells. Although many cells have specific applications, most cells of the body contain exact copies of the original chromosomes in its nucleus, and in the mitochondria surrounding it. Reproduction of many cells goes on throughout life continuously.

DNA is divided into 46 pieces that are contained in the nucleus of every cell in your body. Chromosomes are threadlike strands of DNA that carry the genes containing much of your hereditary information. Your chromosomes come in matched pairs of 23 sets, so every chromosome has a partner, or a "homolog." Each person inherits equally the first 22 pairs, called "autosomes," from his or her parents. For each pair of autosomes, one was inherited from the mother, and one was inherited from the father. During the production of sperm and eggs, each pair of your parent's autosomal DNA, which they also inherited from their parents, was combined to form one new autosome in a process called "recombination." As a result, the autosomes you inherited from your parents are a unique combination of your grandparent's autosomes, who inherited a unique combination of your grandparent's autosomes, who inherited a unique combination of their own parent's autosomes, and so on.

Autosomal DNA makes up the vast majority of DNA in the nucleus of the human cell. It contains almost the entire blueprint, or "genome," for the body. Most of the genes that result in various inherited traits—from eye and hair color to medical susceptibilities—are contained in autosomal DNA.

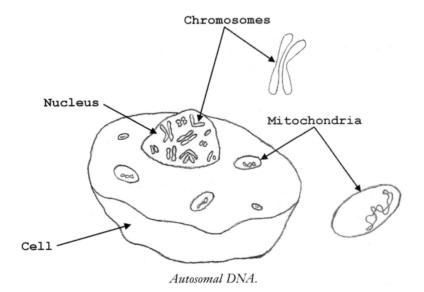

Autosomal DNA.

Autosomal DNA testing is done using these 22 pairs of autosomes *in both males and females.* Anyone can take an autosomal DNA test to look for DNA matches with cousins from all of their ancestral lines.

The 23rd pair, known as the sex chromosomes, determines whether a person is male or female. Females have two X chromosomes; one is inherited from each parent. Males have one X and one Y chromosome. A male's X chromosome is always inherited from his mother, and his Y chromosome is always inherited from his father. Y-DNA testing is done using this Y chromosome in males only.

Y-DNA is passed on directly from father to son for many generations virtually unchanged because there's no input from the female parent. This means every all-male direct descendant of any man living many years ago will have an extremely similar Y-DNA pattern, called a *haplotype.* In addition to possibly connecting ancestry through haplotypes, test results will indicate a haplogroup that provides direction as to the early geographic origins of the paternal line.

DEFINITION

A **haplotype** is a unique combination of "alleles," also called "markers," that are inherited together. The markers are found at different locations on the Y chromosome. All males who are related have similar haplotypes and belong to the same "haplogroup."

Y-DNA tests consist of an examination of markers at various locations on the Y chromosome. These markers are identified by their DYS numbers. (DYS is simply an acronym: D = DNA, Y = Y Chromosome, and S = segment.) In this test, scientists examine Short Tandem Repeats (STRs). Each marker is represented by a number, and this number is actually the number of times a short Y-DNA pattern (or sequence) repeats at a particular location. STRs allow scientists to predict a haplogroup. Single Nucleotide Polymorphisms (SNPs) also occur on the Y chromosome and allow you to confirm an individual's exact haplogroup.

TREE TIPS

If you are unfamiliar with the terminology used (as most of us are!), try one of the following websites:

isogg.org

kerchner.com/glossary.htm

https://genographic.nationalgeographic.com/genographic/lan/en/atlas.html

jogg.info/62/index.html

ysearch.org

Mitochondrial DNA (mtDNA) is passed from a mother to all her children at conception. mtDNA testing is done using this mtDNA in both males and females. Anyone can take an mtDNA test to trace their direct maternal line. The results of an mtDNA test also provide information about your maternal haplogroup. A haplogroup helps you to indentify the general area of where your maternal ancestors lived or migrated as far back as ten thousand years ago.

How to Understand Your Test Results

Most DNA testing companies compare the results of your Y-DNA, mtDNA, and autosomal DNA tests with others in their company's database. Along with the raw data from your DNA test, you also receive a list of your matches from the testing company. To get the most from any DNA test, it's important to contact these individuals to further explore how you may be related. Exchanging information and inquiring about your matches' family histories may lead you to discover important information about your own family.

Y-DNA Results

Your Y-DNA test results appear as a set of numbers that contain the DNA markers tested and their values. Your results appear in three rows. The first row always contains the number of the marker, also referred to as the "locus" (that is, the location on the Y chromosome). The second row always contains the names of the DNA markers, also known as a DYS#. The bottom row always represents a value given to each marker, also known as "alleles," and the results of your test.

Locus	1	2	3	4	5	6	7	8	9	10
DYS#	393	390	19	391	385a	385b	426	388	439	389-1
Alleles	13	24	14	11	12	15	12	12	13	13

It's easy to compare your Y-DNA results to those of another, by simply comparing the values of your alleles with theirs.

Locus	1	2	3	4	5	6	7	8	9	10
DYS#	393	390	19	391	385a	385b	426	388	439	389-1
Kit 1	13	24	14	11	12	15	12	12	13	13
Kit 2	13	24	14	11	11	15	12	12	11	13

The differences between markers are called "steps" (or "mutations"). To calculate, subtract Kit 2 from Kit 1. Here is a calculation based on our example:

385a: 12 – 11 = 1 = one-step mutation

439: 13 – 11 = 2 = two-step mutation

3 = total mutations

The total number of mutations between any two kits is called the *genetic distance*. In the example above, at 10 alleles, kits 1 and 2 have a genetic distance of 3.

LINEAGE LESSONS

The genetic distance, or number of mismatches, between two sets of Y-DNA test results is used to determine the likelihood that two individuals are related, or that an individual belongs to a specific family group in a DNA project. The formula that is generally used to determine the relatedness of two individuals is: Genetic distance of 1 = 25 years, 2 = 50 years, etc.

Your relationship with any individuals you match with on your Y-DNA test is determined by both the number of tested alleles and the genetic distance.

On a Y-DNA 37 marker test, these individuals would be considered a 34/37 match. This is calculated by:

37 = Total # of markers
<u>–3</u> = Total number of mutations (genetic distance)
34

If the individual you match on your Y-DNA test shares the same (or a similar) surname, it's "expected" you're most likely related. If your surnames are different, to rule out false positives and to prove your relationship, you (and this individual) need to upgrade to a 67- or 111-marker test. A 111-marker kit can provide more definitive results by not only confirming whether two individuals are related, but also providing a prediction of how many generations back you share a common ancestor.

LINEAGE LESSONS

The more markers you match with an individual on a Y-DNA test, the more likely it is you're related. The more differences, or mismatches, the less likely you're related.

mtDNA Results

Your mtDNA test results provide you with a list of matches who are related to you through your direct maternal line. Depending on the type of mtDNA test you ordered, your test may provide information to assist you in discovering your relationship to those you match.

There are three main levels of testing on mtDNA:

- **Matching on HVR1 test:** Indicates you have a 50 percent chance of sharing a common ancestor within 52 generations, or 1,300 years.

- **Matching on HVR1 and HVR2 tests:** Indicates you have a 50 percent chance of sharing a common ancestor within 28 generations, or 700 years.

- **Matching on an mtDNA Full Mitochondrial Sequence (FMS) test:** Indicates you have a 50 percent chance of sharing a common ancestor within 5 generations, or 125 years.

Autosomal DNA Results

Your autosomal DNA test results provide you with a list of matches. You can view your autosomal DNA test results in either of two ways. The first is in the form of raw data. The second is by viewing a list of those with whom you match and a predicted relationship, which is based on the length and number of identical segments of DNA you share with them. If you match with someone on your autosomal DNA test, it indicates you share a common ancestor. To match with someone on your test, you must share the exact same DNA, and therefore, a set of grandparents, great-grandparents, and so on.

Your probability of matching with cousins who have also tested is as follows:

99 percent—2nd cousins, or closer

90 percent—3rd cousins, or closer

50 percent—4th cousins, or closer

10 percent—5th cousins, or closer

<2 percent—6th or more distant cousins

While an autosomal DNA test is not a legal paternity or sibling test, it can indicate a close relationship for two individuals who've taken the test. If both you and your mother (or father) take this test, your results will indicate a parent–child relationship. An autosomal DNA test can also identify the following relationships: sibling, half-sibling, grandparent, and aunt or uncle.

Sharing Your Results

You have the option of uploading your results to public databases that make it possible to compare your results to others who ordered the same tests as you, from other testing companies. Family Tree DNA hosts a public Y-DNA database called "Y-Search," and a public mtDNA database called "Mitosearch."

GEDmatch.com is a public database that allows you to upload your autosomal DNA test results, and compare them with those who have tested at other companies. While this website is not sponsored by any of the testing companies, anyone can use it for free.

> **TREE TIPS**
>
> Unlike Family Tree DNA's public databases, you cannot delete your results from GEDmatch.com's database. If you upload your results, or your GEDCOM, to their database and later change your mind, simply upload a blank file or set of data to the site. To do this, create a file, delete all of your information, and upload your file to their site.

What Started All This Fuss?

Scientists first identified DNA in the 1950s. Over the following 20 years, it became clear to scientists that DNA carries the blueprint for all life on earth.

By the 1980s, it had been established in the scientific community and beyond that DNA could be used very effectively for identification purposes, leading to widespread use in the forensics and criminology fields. Only very small samples are needed for testing, with sources varying from a body to traces left by it, in what is referred to as "genetic fingerprints." DNA paternal testing soon followed, effectively proving paternity far beyond the limits of earlier blood tests. A major achievement occurred in the early 1990s, with the exhumation and identification by DNA testing of the remains of members of Russia's royal Romanoff family.

A genetics study at the University of Arizona led to the formation of Family Tree DNA in early 2000. A study by Dr. Brian Sykes of Oxford around the same time led to the formation of Oxford Ancestors. These companies were the first to offer commercial testing services to the public. Although a major turning point, these early tests were based on a small number of markers, and thus were somewhat limited in scope. This changed, however, as they expanded their tests to include more alleles. DNA testing from the Y Chromosome (Y-DNA) clearly became the most advantageous for genealogical purposes. Surname projects became very common, providing the lion's share of testing. Projects were usually started by enthusiastic researchers from various families, intent on identifying the early origins of their surname, possible connections, individual branches descended from a common ancestor, and unrelated groups with the same surnames.

By 2002, Y-DNA testing expanded quickly to 25 markers. Other firms also began offering tests. Thirty-seven- and 42-marker tests were available by 2004, creating a much higher degree of certainty than possible previously. Now 67- and 111-marker tests are available.

Privacy Issues

In addition to confirming paternal relationships, DNA testing is capable of disproving a father–son relationship. A test can easily establish brothers aren't truly brothers, or paternal cousins aren't really cousins. Such results disprove family connections, effectively knocking down a very sizable portion of a family tree that was created painstakingly. Be prepared. What is referred to as a *nonpaternity event* is more common than one may suppose. You must enter into DNA testing with an open mind, in the spirit of finding out what is really so, even if it means disregarding years of research on what you thought was your family. If you aren't prepared for unexpected results, don't take the test.

> **DEFINITION**
>
> A **nonpaternity event** may be the result of illegitimacy, adoption, raising an orphaned child without benefit of a formal adoption, and so on. The rate of such nonpaternity events is generally estimated at about 2–3 percent per generation.

What Can You Expect from the Project?

- Hopefully you will confirm relationships—or at least stop going down the wrong path and be guided to a different lineage suggested by the DNA results.

- You may be able to direct your research to a specific area or time period.

- Your test may match someone with a variant spelling you hadn't considered.

DNA testing is one tool in the many genealogists use. It can't perform miracles, but it can open new avenues of research.

The Least You Need to Know

- Only males with an unbroken father-to-son descent can participate in Y-DNA testing.
- A minimum of 37 markers are needed on a Y-DNA test to sufficiently compare results with other tests (though 67 markers are preferable).
- A Mitochondrial DNA (mtDNA) test is especially useful when the trail of a female ancestor stops but you know more about a sister.
- An autosomal test such as Relative Finder or Family Finder can help in predicting the degree of relationship between two persons.
- Participating in a group project is normally less expensive, and has the advantage of a large database of tests for comparison.
- DNA test results are more meaningful when coupled with traditional genealogical research.

After the Basics— What?

In This Chapter

- Immigration and naturalization records
- Homestead records
- Passport records
- Tax records
- Native American research methods
- Ethnic records

Previous chapters discussed the fundamental records for your research: federal population census records, deeds, wills, estates, Revolutionary War records, Civil War records, and bounty land records. Those alone will keep you busy for some time. As you gain experience, you'll learn to think in terms of, "What other records will likely tell me more?"

The records you'll need to solve your advanced research problems are sometimes somewhat obscure. There are good reasons to utilize some records less often: many are unindexed, their genealogical value may be unknown, or they may be more difficult to locate. They may not be filmed or digitized, so they aren't available unless you travel or hire a researcher. Nonetheless, being aware that certain records exist may be your key to solving a seemingly impossible problem.

Our Immigrant Ancestors

Your father tells of hearing his grandfather talk about the experience he had as a young man, barely 18 years old, leaving Germany and arriving at a U.S. port with almost no funds, but an unbounded excitement and dreams of finding his fortune in a new country. Your mother says her great-grandparents left Ireland during the potato famine, destitute but convinced they had to leave family and friends behind. Your aunt thinks an ancestor came over on the Mayflower.

We yearn to know more about our ancestors' arrival into this country—the name of the ship, the date they arrived, what their experiences may have been en route.

A good starting place is Chapter 24, "American Aids to Finding the Home of the Immigrant Ancestor," in Greenwood's *The Researcher's Guide to American Genealogy* (3rd edition). Here you'll find a discussion of customs passenger lists, immigration passenger lists, customs lists of aliens, and much more. An extensive bibliography follows the chapter.

Another great source is Chapter 9, "Immigration Records" in the 3rd edition of Szucs and Luebking's *The Source* (this chapter was written by Loretto Dennis Szucs, FUGA, Kory L. Meyerink, MLS, AG, FUGA, and Marian Smith). This will take you through all you need to know to start your search, and includes many illustrations and charts to help you visualize the processes.

If your family came very early into the New England area, you'll want to study the seven volumes published in the Great Migration Study Project. The goal of this ambitious project (under the direction of Robert Charles Anderson, MA, FASG) was to compile comprehensive genealogical and biographical accounts of every person who settled in New England between 1620 and 1640. Sketches from volumes 1–3 are available at ancestry.com. Read more about this at greatmigration.org.

Don't overlook the U.S. Citizenship and Immigration Services (USCIS) website at uscis.gov/portal/site/uscis, or the informative article by Marian L. Smith, "INS— U.S. Immigration & Naturalization Service History" (the INS is now the USCIS), available online at uscitizenship.info/ins-usimmigration-insoverview.html.

On the High Seas

Immigrants usually arrived by ship, and may have come into any number of port cities: New York, Boston, Baltimore, Philadelphia, New Orleans, Key West, Mobile, Charleston, Savannah, New Bedford, New Haven, Providence, San Francisco,

Galveston, Seattle, Port Townsend (Washington), Gulfport, Pascagoula (Mississippi), and others. You can also use your favorite search engine to look for passenger lists in a smaller area. For example, insert "San Francisco passenger lists" and you'll come up with several websites including sfgenealogy.com/sf/sfdata.htm. Another useful website is immigrantships.net.

LINEAGE LESSONS

Indexed ships' passenger lists are increasingly available online. Thousands arriving in New York between 1892 and 1924 passed through Ellis Island. Volunteers for The Church of Jesus Christ of Latter-day Saints indexed the National Archives' microfilms of New York passenger lists for that time period. The indexes are searchable on ellisisland.org. A more comprehensive search engine for these records, developed by Dr. Stephen P. Morse, is at jewishgen.org/databases/eidb; it offers better control of searches. Ancestry.com has indexes for passengers on ships from foreign ports to New York, 1851–1891. Their indexes are accessible by subscription only, but many libraries subscribe and offer access to their patrons.

Passenger arrival lists were created originally for customs; they contain less information than those created as a result of immigration laws beginning in 1882. Because so many of the passenger lists aren't indexed, to use them you need to know the ship or date of arrival, often the very information you seek. Nonetheless, it's worthwhile to learn about these records. Even the early customs lists are useful because they usually give the age, occupation, country of origin, and place of intended settlement for each passenger. You may find the whole family traveling together, perhaps accompanied by other relatives or friends. Be careful, however, of accepting the information on these lists without further confirmation. On a ship sailing from Liverpool in 1858, the country of origin for all the passengers is listed as Great Britain when, in fact, they were from many European countries.

To search thousands of ships' passenger lists, go to stevemorse.org/ellis/passengers. php?mode=ny and complete the search form. Within seconds you may find the ancestors you're seeking. If so, clicking on the entry will direct you to any available associated online images.

A good article by Myra Vanderpool Gormley, CG, "What You'll Learn from Passenger Lists," is at genealogy.com/8_mgpal.html. It includes a number of links to other related topics.

Records amassed after the passage of immigration laws are more useful. Immigrants were asked many questions by customs officers, such as whether they had been in the

United States before, and if they were going to join a relative (and if so, the relative's name, address, and relationship). Later, the lists included a physical description and the name and address of the nearest relative in the immigrant's home country.

Consult Michael Tepper's *American Passenger Arrival Records* or Greenwood's *Research in American Genealogy* for good discussions of passenger lists. Also helpful is John P. Colletta's *They Came in Ships*.

Swearing Allegiance—The Naturalization Process

Naturalization records are often neglected as a source of information because they're difficult to locate. Continually tinkering with laws governing naturalization will likely result in confusion as to the naturalization process, where the records are, and what they contain. For many years, any court (federal, state, county, or local) could accept the declaration of intention to become a citizen and grant citizenship once the individual had fulfilled the requirements. Since 1906, naturalization proceedings have taken place in federal courts. Christina K. Schaefer's *Guide to Naturalization Records of the United States* is a county-by-county guide and, though not nearly complete, may help you locate these important records.

Because of continual changes in the law, no blanket statements covering naturalization apply to all situations. Generally, the citizenship process began with the *declaration of intention,* or *first paper,* in which the alien states a desire to become a citizen and renounces allegiance to his native country. When the prevailing residency requirement was fulfilled, the alien went back to court, petitioned for citizenship, and took an oath of allegiance to the United States. The first paper was usually filed in a court near the alien's residence, although any court of record could accept it until 1906. The petition and final certificate could be in a completely different state if the alien moved. In some time periods, the alien's wife and children automatically became naturalized when the husband received his certificate. After 1906, records pertaining to naturalization are with the USCIS and in the courts.

DEFINITION

Naturalization is the legal process for an individual of foreign birth to become a citizen of the United States.

The **declaration of intention,** or **first paper,** is a sworn statement by an alien that he intends to become a citizen.

Some sources:

- Marian L. Smith, "Women and Naturalization, ca. 1802–1940." *Prologue: Quarterly of the National Archives*, Vol. 30, No. 2 (Summer 1998): 146–153. Available at archives.gov/publications/prologue/1998/summer/women-and-naturalization-1.html.

- The National Archives website on the subject is available at archives.gov/research/naturalization/naturalization.html.

To find more websites, type "naturalization" into your browser. Try refining it further with words such as "images," "registers," the name of a state, and so on. Also use the subscription websites to locate databases they have indexed or digitized.

Microfilmed naturalization records are listed in the microfilm publication catalog of the National Archives, *Immigrant & Passenger Arrivals: A Select Catalog of National Archives Microfilm Publications.* See also NARA's *The Guide to Genealogical Records in the National Archives* and *The Archives: A Guide to the Field Branches of the National Archives.* John J. Newman's *American Naturalization Processes and Procedures 1790–1985* will add even more information.

Passport Records

Although passports to foreign countries weren't required until World War I (except for a brief period during the Civil War), many individuals obtained them as a little extra protection. This was especially true for male immigrants who were afraid of being conscripted into the military service in their country of origin if they journeyed there. Immigrants to the United States did return to the "Old Country" to visit relatives, marry, take a child to see his or her grandparents, or bring a family to the United States. Later passport records are more complete, but you can usually get some personal information from earlier ones.

Read the important information at archives.gov/research/passport/index.html. This website also has links to full descriptions of two important microfilm publication series, the *Register and Indexes for Passport Applications 1810–1906*, M1371, and *Passport Applications 1795–1905*, M1372. See also the large collection of passport records at ancestry.com.

Consider also that passports were needed in early times to pass through Indian territory and other "foreign" territories within the United States. See Mary G. Bryan, *Passports Issued by Governors of Georgia, 1785 to 1820* published in the National

Genealogical Society quarterly in 1977 and indexed at http://freepages.genealogy. rootsweb.ancestry.com/~texlance/earlyresidents/passportsindex.htm. See also Dorothy Williams Potter, *Passports of Southeastern Pioneers, 1770-1823: Indian, Spanish and other Land Passports for Tennessee, Kentucky, Georgia, Mississippi, Virginia, North and South Carolina* (Baltimore: Gateway Press, 1982).

And More, and More, and More

The array of records available with information about your ancestors is immense. We can't discuss them all. But here are a few important ones to learn about.

The Homestead Records

Among the many *federal land* records are the homestead records. The Homestead Act of 1862 was instrumental in settling the West. Many of our ancestors submitted papers and fulfilled the requirements to claim the allotted 160 acres. Any citizen (or person who had filed papers to become a citizen) over the age of 21 and who had not borne arms against the United States could claim 160 acres of unoccupied, available federal land. The individual had to live on the land for five years while cultivating it and building a home. Women as well as men were homesteaders, so don't overlook your female ancestors.

DEFINITION

Federal land is land owned by the federal government from the end of the Revolutionary War; that is, this land was not owned by states or under private ownership.

The case files for the homestead lands vary, but nearly all have valuable genealogical information, such as the age and address of the person applying for the land, family members, descriptions of the land, house, crops, and witness testimony. For a naturalized citizen, you'll find information about the immigration, such as the date and port of arrival, and the date and place of naturalization. Some files even have copies of discharge from Union service in the Civil War because a subsequent act gave special privileges to those veterans.

Homestead files are in two series: one for completed land entries and one for canceled land entries. Both files are useful. The completed land entry files are in the custody of the National Archives in Washington, D.C., in Record Group 49. The canceled land entry papers are scattered among a variety of repositories, but they can yield clues as to why the requirements weren't fulfilled by the applicants.

For help in understanding and locating homestead and other land records, consult Szucs and Luebking's *The Source*, Greenwood's *Research Guide to American Genealogy*, and Wade Hone's *Land and Property Research in the United States*.

> **TREE TIPS**
>
> There were other acts, earlier and later, pertaining to acquisition of federal land by private individuals. There are case files for cash entry sales, preemption sales, timber culture, desert land, and others. All of these are housed in Record Group 49 at the National Archives in Washington, D.C. These records don't usually contain as much personal information as the homestead files but, nonetheless, they're worthwhile.
>
> The records of the U.S. patents, the document that actually transferred federal land to an applicant after all steps had been successfully completed, is accessible at glorecords.blm.gov. For detailed instructions on how to use this website, see Christine Rose's *Military Bounty Land 1776-1855,* Chapter 6. It includes images and step-by-step instructions on how to use and search the Bureau of Land Management website.

Taxing Matters

Tax records are very important to genealogists. There were poll taxes, a tax on all free white males in a community over a specified age, and property taxes, both real and personal. Personal property tax records can document the existence of ancestors who didn't own land. They verify that an individual was in a particular place at a particular time. If the person remained there, he should appear on the tax rolls year after year. Property tax rolls reflect the acquisition or divestiture of land. The tax information may guide you to more records. If an individual disappears from the tax rolls, you'll know he reached an age (or condition, such as blindness or poverty) of tax exemption, died, or left the area. Any one of those reasons is valuable to your research.

It's best to use tax records not in isolation, but in a series. If you find your ancestor on a tax list, follow him year by year back to his first appearance and forward to the year he disappeared from the roll. There may be some gaps in the existing records, but try to follow your ancestor throughout the rolls. Occasionally, abstracts of tax records are published and indexed, but original tax records generally aren't indexed.

The Government Gets Into the Taxing Act

Before there was a federal income tax, the federal government reached into your ancestors' pockets several times. The surviving tax lists contain little or no genealogical information, but they can give you a glimpse of the economic status of your ancestor. The first direct tax was a 1798 tax on real property and slaves. The National Archives has the Pennsylvania lists (*United States Direct Tax of 1798: Tax Lists for the State of Pennsylvania*, microfilm publication M372), which are organized in a complex geographic division. What a wonderful resource it is! For individuals on the lists, you'll find the size and construction material of their houses, the number of windows and lights in the houses, and how many stories they were. Acreages, outbuildings (such as a milk house or distillery), and other tantalizing facts are there. Sometimes the list gives adjoining landowners, enabling you to sort out men of the same name. The other surviving lists (Connecticut, Delaware, District of Columbia, Maine, Maryland, Massachusetts, New Hampshire, Rhode Island, Tennessee, and Vermont) are in various repositories and are well worth hunting for if you have ancestors in those areas at that time.

Easier to use are the Internal Revenue Assessment Lists for the Civil War period. To pay for the Civil War, taxes were levied on various businesses and licenses. Carriages, yachts, billiard tables, and gold and silver plates were taxed as luxury items. In 1865, one of my collateral relatives was taxed $10 each for his licenses as a claims agent, insurance agent, real estate agent, and retail dealer. He was taxed $25 as a retail liquor dealer, and assessed $1.00 each for his carriage and two gold watches, and $4.00 for his piano. No genealogical information here, but you can get an idea of his economic status from the tax list. As with most records, it helps to read the background information on whatever taxes you are studying. An informative article is "Income Tax Records of the Civil War Years," by Cynthia G. Fox. Originally published in *Prologue*, it's now available online at archives.gov/publications/prologue/1986/winter/civil-war-tax-records.html.

Native American Research

Researching Native American records requires some attention to the focus. Are you trying to verify that you do have Native American ancestors? Or have you already confirmed that, and are interested in the particular tribe such as Shawnees, Creek, or Cheyenne? Or perhaps you're consumed with an interest in a specific Native American leader such as Tecumseh, Chief Joseph, or Geronimo. Your approach to the research will depend upon your focus.

Start your search with the same techniques as you would use for any other quest for ancestry. Talk with your family, listen to the traditions, and scour through the family's mementos for clues. Then branch out to court records, census records, military records, and others. You'll be surprised as you pursue the family to discover there are more records than you might at first imagine.

Go to genealogybranches.com/nativeamericans.html for a large number of links to Native American records and finding aids. You'll find you can access records of Native Americans mustered into the service of the United States in the War of 1812, U.S. Indian Census Schedules, the Chapman Roll of Eastern Cherokees 1851, and much more. Do you suspect your ancestors were married in the latter part of the nineteenth century in the Chickasaw Nation? Perhaps you'll find them in the 1855–1907 online database at chickasawhistory.com/m_index.htm.

In earlier times, descendants of Native American ancestry were reluctant to reveal or talk about their heritage. Now, times have changed and those who suspect they have some Native American ancestry in their family try diligently to document the connection.

Those with an interest in this subject should read Chapter 19, "Native American Research," by Curt B. Witcher in Szucs and Luebking's *The Source* (3rd edition). This chapter will provide you with invaluable assistance, including a list of references at the end of the chapter.

Also be sure to study the website at archives.gov/research/native-americans. This National Archives site has links you'll want to examine. "Enumeration of Seminole Indians in Florida, 1880–1940, Indian Bounty Land Applications," an article from *Prologue* magazine, is among the offerings, along with others.

The National Archives branch at Fort Worth, Texas (archives.gov/southwest/public/research.html), holds many items of interest, some not found in other Archives' branches.

Don't neglect the excellent Chapter 11 in *Genealogical Research in the National Archives of the United States* (3rd edition), "Records of Native Americans." After reading it you'll no longer believe that there's a paucity of Native American records!

There are many excellent records involving Native Americans on the Internet. A few:

- archives.gov/research/arc/native-americans-guion-miller.html (Includes an index to the applications submitted for the Eastern Cherokee Roll of 1909 [Guion Miller Roll] and description information)

- http://memory.loc.gov/ammem/browse/ListSome.php?category=Native%20 American%20History

- fold3.com (For several Indian-related databases)

- http://genealogicalstudies.com (For information on a course on Native American genealogy)

- familysearch.org (Go to the homepage, click at the bottom on "Free Courses," and follow links to Ethnic groups and then to Indians—you will find a video by Janice Schulz on Cherokee research. By using the same link but scrolling to Native races instead of Indians you will find a video by Kathy Huber of the Tulsa City-County library on the use of the Dawes rolls.)

These are just a few of the many websites that will be useful in your search.

Ethnic Records—Other Countries

If your ancestor was a member of a particular ethnic group, your research will be enhanced by seeking special resources.

New books by ethnic specialists appear regularly; watch for those applicable to your interest. Search for the collections of religious denominations and periodicals devoted to various nationalities. Numerous federal government records exist that pertain to particular groups.

Genealogists who specialize in ethnic groups, whether racial, national, religious, linguistic, or cultural, often lecture at national conferences. Many presentations are on tape and are available for purchase through jamb-inc.com.

Research in England, Ireland, and Northern Ireland

Some helpful sites with which to start are the following:

- british-genealogy.com/resources-guides/actual-resources.html (many links and information)

- bi-gen.blogspot.com (a blog)

Also read Chapter 15 of Szucs and Luebking's *The Source* (3rd edition), "Colonial English Research" (by Robert Charles Anderson, MA, FASG). The author examines English research in the New England colonies and several southern colonies.

At cyndislist.com/uk, many links help with searches in the United Kingdom and Ireland: general resources, libraries, births, deaths, and the list goes on and on. See also ukgenealogy.co.uk/england.htm for links to county research in England, societies, and more. Read also "Research in England" at genealogy.com/32_donna.html.

For a variety of Scottish topics, try scotlandspeople.gov.uk. To reach the various UK societies, use your browser and insert the country, such as "England Genealogy Society," etc.

There is much more—in your browser enter "England genealogy," or "Scotland genealogy," or whatever you desire, and you'll be amazed.

Those searching for Irish ancestors can try www.irishorigins.com/Welcome.aspx.

German Research

A good starting point for research into German ancestry is the website German Roots (germanroots.com/germandata.html). This site will lead you to a number of links, including links and indexes to German Emigration Records. Here you'll find links to records of Baden-Württemberg, Brandenburg, Hesse [Hessen], and more. Use your browser to search specific jurisdictions such as Hesse, and others.

Look also at cyndislist.com/germany for links to how-to articles, societies, information on Palatine Germans, maps, and others.

French Research

Those who are searching for French families will find a rich source of information on the Internet. Use cyndislist.com/france to get started.

The National Archives of France has some online materials, but researchers do very little genealogical research at their National Archives. One goes to the departmental and municipal archives for genealogical research. More than half of the 101 French departmental archives and dozens of municipal archives have put original records online (the great majority for free). They include parish registers (back to the beginning), civil registrations (with indexes), census records, cadastral records, military conscriptions, old newspapers, and more.

Try the website at afgs.org/genepges.html for many links that can assist you. If you read French you can go to a map at francegenweb.org/~archives/archivesgenweb/?id=carte for information on which archives are online and which are in progress.

Brigham Young University and FamilySearch.org offer French language courses.

Hispanic Research

Start your research with Chapter 17 in Szucs and Luebking's *The Source* (3rd edition), titled "Hispanic Research" (written by George R. Ryskamp, J.D, AG). You might also look at Ryskamp's book *Finding Your Hispanic Roots.* To locate websites for a particular country, insert "Hispanic Genealogy" into your browser and select the links pertaining to the appropriate country, such as Spain, Mexico, or Colombia.

Check also cyndislist.com, which categorizes Hispanic research into three groups: "The Caribbean/West Indies," "South and Central America," and "Mexico." Click on any of those to get links to many websites. Examine also familysearch.org/learn/researchcourses for their course on Mexican genealogy.

Italian Research

The many links at angelfire.com/ok3/pearlsofwisdom/boards.html will get your Italian research going. An Italian dictionary and some grammar tips on reading the language are found at familysearch.org/Eng/Search/rg/Guide/WLItalia.ASP.

An informative monthly newsletter published by the Italian Genealogy Group (italiangen.org/Default.htm) gives tips, websites, firsthand accounts of members' trips to Italy, and much more. Another organization, POINT (Pursuing Our Italian Names Together) also publishes a magazine and offers other services to their members. See their website at point-pointers.net/home.html.

FamilySearch.org lists at least five free videos/slides devoted to Italian Research at https://www.familysearch.org/learningcenter/home.html.

Trafford R. Cole's *Italian Genealogical Records* published in 1995 (Salt Lake City, UT: Ancestry Incorporated, 2009) is an important guide to researching these families. See also the second edition of John Philip Colletta's *Finding Italian Roots* published in 2009 (Baltimore, MD: Genealogical Publishing Company).

Your browser will lead you to many other links to enhance the research on your Italian ancestors. Learning how to locate their records and studying the culture of the times in which your family lived there will add considerable interest to your compilation of your family lines.

Asian Research

At www.cyndislist.com, click on "Categories" and then "Asia & The Pacific." From there you'll see sites listed from many different countries and types of records and even libraries, archives, and museums of different countries. Historical maps, mailing lists, internees—the avenues of research and historical prospective are endless through these sites.

An upclose look at the internment of Asians can be viewed in videos and interviews at www.aiisf.org, the website for Angel Island Immigration Station.

Still More Federal Records

The scope of federal records with genealogical information is mind-boggling. Just glancing through the *Guide to Genealogical Research in the National Archives* will start you thinking. Some of the records of the federal government will be useful to you only after you've done a great deal of research in other sources. You can use others early on in your search if you have access to them. This selected list (in no particular order) gives you an inkling of the kinds of records that might be useful at some point in your research:

- Bureau of Indian Affairs
- Bureau of Refugees, Freedmen, and Abandoned Lands
- Southern Claims Commission
- U.S. Marshals
- U.S. Coast Guard
- Seamen's Protection Certificates
- Department of the Interior
- U.S. Treasury
- Bankruptcy
- Birth, Death, and Marriages at U.S. Army Posts
- Diplomatic Records
- Records of the Continental Congress
- Veterans Homes
- Amnesty and Pardon Records
- Japanese Internments

Archival research is somewhat different from other kinds of research. Because of the vastness of the federal records, there are no overall indexes. Records are organized as they were when the originating federal agency used them. The biggest obstacle to using federal records is that many of them aren't on microfilm. You'll have to go to the National Archives or one of its branches, or hire a researcher to look at the records for you.

The National Archives has published many useful guides to its records. There are microfilm publication catalogs, preliminary inventory lists created by the archivists, and the three-volume Robert B. Matchette, et al., *Guide to Federal Records in the National Archives of the United States.* The *Guide* is available online at archives.gov/ research/guide-fed-records. Check at archives.gov/research_room/genealogy for many pointers for using federal records.

Potpourri of Other Records and Sources

You may have to be creative in your thinking to discover your ancestors in other records. Did someone work for the railroad? There are records. Maybe they came into the country through Canada or Mexico; find pertinent border-crossing records. High schools and universities often have student evaluations more than 100 years old. Prisons, too, keep records and not just for the inmates; my third great-grandfather was a guard at an Illinois penitentiary and his file was extant in the 1980s even though he retired in 1883. Look for connections to fraternal organizations and not just for men. Women belonged to auxiliary groups of the Masons, the International Order of Odd Fellows, and others. Grandmother may have belonged to Royal Neighbors of America, and their headquarters can tell you when she cashed in her insurance policy. Biographies exist for even less-famous politicians at the state and county level. Could your ancestor have been on an orphan train? Your ancestor may be mentioned in someone else's records. Physicians and storekeepers kept account books that can detail services rendered to your ancestor.

The minutiae of your ancestors' lives differentiates them and makes them more than just a name on a pedigree chart. It's up to you to accumulate the minutiae wherever you find it.

Modern technology makes it possible and practical for companies to amass huge collections of unrelated data. Some companies claim you can now find your ancestors with a few clicks of a computer mouse. We've heard ads for online sites promising to find your family in five minutes. In that raw data, you may find individuals with the same names as your ancestors, but no proof they're the individuals you're seeking. Good genealogy requires the researcher to go beyond indexes and undocumented lists. There are some shortcuts in genealogy, but there are no shortcuts to proving you've discovered *your* ancestors. Technology continues to provide new tools to cut your research time, but there's no substitute for the systematic, thorough research necessary to prove a line.

It's beyond the scope of any book to give you in-depth coverage of every record available for genealogical research. Be on the lookout for new-to-you sources of information. Rich sources of genealogical information are everywhere, mostly in records not created with genealogy in mind. Weave Internet research into your overall plans, but don't depend on it to the exclusion of all other sources.

Your enthusiasm will build as you discover the continuity of history told through the fascinating stories of your ancestors. Whether you research in a traditional library or the digital world, healthy skepticism is important when evaluating your finds. Remember most of all that tracing a family is fun. It strengthens family ties, reconnects lost branches, and forges new friendships. Enjoy the journey.

The Least You Need to Know

- Rich sources may not be on microfilm or digitized, but remain in their original form at the custodial repository.
- Unique records can be in county, state, or national repositories; in university special collections; and in private collections.
- Sources for immigration, naturalization, Native American research, ethnic groups, and much more abound on the Internet. Also check libraries for published books.
- Read widely to learn about records that deliver the goods on your ancestors. Trust, but verify.

Glossary

abstract   Summary of a document that retains every important or pertinent detail.

administration   Decedent's estate for which there is no will.

administrator   Person appointed by the court to handle the estate of one who died without a will.

affidavit   Written declaration made under oath before a notary public or other authorized official.

age of majority   Legal age of adulthood; varies by area.

ancestor   Person from whom one is descended.

antebellum   Comes from Latin words meaning "before war." In American history, antebellum is used generally to designate the period before the U.S. Civil War, specifically the early- to mid-nineteenth century until 1860.

ascendant chart   Starts with an individual and moves back through each generation of his or her ancestors.

banns   Announcement, usually in church, of an intended marriage.

beneficiary   One who benefits from a provision made by another, usually in a will.

Black codes   Laws established during the antebellum period which specifically restricted the rights of African Americans. These laws existed in the north and south.

bookmark or **favorites**   A shortcut list to users' preferred websites and web pages.

bounty land   Federal land awarded by the federal government for military service.

buccal swab   Specimen taken by brushing inside the cheek or mouth cavity.

bulletin board   Online gathering for posting messages on a common topic.

census   Official count that often includes related information for government planning; a *population* census is a count of people residing within a designated area.

Christian name   First (and middle) name; the name given to a child at birth or baptism. Also called *given name*.

citation   Formal notation of the source of information.

cite   Call attention to the proof or source of information.

Civil War census   Special 1890 federal census taken of Union veterans of the Civil War.

closed stacks   Refers to library books that patrons must submit a request to the librarian to examine. See also *open stacks*.

collateral relative   Someone with whom you share a common ancestor, but who is not a direct line.

consideration   Asset given by the buyer of a property to the seller of the property; may or may not be monetary.

consolidated index   Index combining the indexes of several sources.

county clerk   Clerk of the court handling particular documents and transactions in a courthouse.

county seat   Town or city that is the administrative center for a county.

Daguerreotype or **tintype**   Photographic processes in which images appear on light-sensitive, silver-coated, metallic plates or directly on an iron plate varnished with a thin sensitized film.

decedent   Deceased person.

declaration of intention   Sworn statement by an alien that he intends to become a citizen. See also *first paper*.

deed   Legal document used to transfer title.

descendant chart   Chart that starts with an individual and comes down through the generations listing the individual's descendants.

digitize   To convert data to numerical form for use via a computer.

directory   List of categories, usually alphabetical, with links to websites.

DNA   Deoxyribonucleic acid, the molecule carrying the genetic code for an individual.

DNA profile   Set of specific values for genetic markers inherited as a unit. Also called *haplotype*.

double dating   Practice developed because of calendar changes: dates from January 1 to 25 March before 1752 are shown with both the old calendar year and the new calendar year; for example, January 23, 1749/50. The first indicates the year under the old calendar, the second the year under the new calendar.

dower   Portion of an estate allotted by law for a widow.

drop chart   Connects two people, one of an earlier generation and one of a later generation, showing their link generation by generation.

emigrant   Person who departs from one country to establish permanent residence in another. See also *immigrant*.

Enumeration District   Area assigned to census taker for counting.

enumerator   One who makes a list, usually for census or tax purposes.

estate   The whole of one's possessions, especially the property and debts left by a person at time of death (or the possessions of a minor, incompetent person, or other in need of protection by a court process).

et al   Shortened from *et alii*, Latin meaning "and others."

et ux   Shortened from *et uxor*, Latin meaning "and wife."

et vir   Latin term meaning "and husband."

evidence   Information offered as proof of a lineage or relationship, or other fact.

executor   Person appointed by the testator to handle an estate after the testator's death.

family association   Membership group descended from a particular ancestor or group of ancestors.

family group sheet   Form used to record information on a family unit.

family traditions   Stories handed down from generation to generation, usually by word of mouth.

first paper   A sworn statement by an alien that he or she intends to become a citizen. See also *declaration of intention*.

forum   An online meeting place where messages can be exchanged.

gazetteer   A geographical dictionary.

Gedcom   A standardized method of formatting your family tree data into a text file that can be easily read and converted by any genealogy software program.

genealogy   The account—or history of descent—of a person, or a study of a person's family.

given name   First (and middle) name; the name given to a child at birth or baptism. Also called *Christian name*.

grantee   Buyer of property.

grantor   Seller of property.

haplogroup   Similar patterned and related descendant haplotypes that are generally indicative of common geographic origins.

haplotype   Set of certain values for genetic markers inherited as a unit. Also called *DNA profile*.

infant   Person who has not reached the age of majority.

immigrant   Person who comes to a country from another to establish permanent residence. See also *emigrant*.

indentured servant   One who enters into a contract binding oneself into the service of another for a specified term, usually in exchange for passage or entrance into a country.

instant (inst.)   Indicates the date referred to is in the same month as a date mentioned previously.

interlibrary loan   System in which one library lends a book or microfilm to another library for use by a patron.

intestate   One who dies without a valid will.

lineage society   Organization whose members are descended directly from particular ancestors (such as Revolutionary War veterans or tavern-keepers).

LSASE   Long self-addressed stamped envelope (usually a #10 envelope).

mailing list   An automatic email system from which subscribers receive all postings.

manumission   To willfully free one from slavery or bondage, as opposed to "emancipation," which is the end of legalized slavery.

markers   Specific identifiable and measurable positions on a DNA ladder.

maternal ancestor   Ancestor on the mother's side of the family.

memorabilia   Items of significance to a family or a person.

micropublication   Microfilm or microfiche, or a series of microfilm or microfiche.

mitochondrial DNA or **mtDNA**   The power pack of the cell containing traits inherited only from the mother; used to establish maternal relationships from the distant past.

mortality schedule   Special federal census records that list those deceased within a certain prescribed period (usually June 1 of the previous year to May 31 of the current year); sometimes accompany federal census records.

mortgage   Pledge to repay money borrowed.

mug book   Slang term for county or town history, containing—among other items—biographies. Often accompanied by pen sketches or photographs.

muster roll   List of all personnel (officers and enlisted) in a military unit; includes name, rank, absences, money owed, and more.

NGSQ numbering   Genealogical numbering system, a modified *Register numbering* system. Instead of giving an Arabic number to each child, in the NGSQ system all children are given an Arabic sequential number. A "+" is inserted before each of the children's Arabic numeral to designate those carried forward.

naturalization   Legal process an individual of foreign birth must complete to establish citizenship.

necrology   List of persons who died within a certain time frame; collection of obituaries.

online catalog   A database that can be accessed by computer via the Internet.

open stacks   Refers to library books that patrons may examine freely. See also *closed stacks*.

original material   Loose papers, letters, photographs, diaries, and other items included in manuscript collections.

orphan   In early times, an infant under the age of majority whose parent or parents were deceased, commonly used when the father was deceased but the mother was living (though in some instances the mother was the deceased parent). Now refers usually to one whose biological mother and father are deceased.

parent county   County from which a new or present-day county was formed.

paternal ancestor   Ancestor on the father's side of the family.

patronymics   Names derived from a father's name, usually based on the father's given name or from the paternal side of the family.

pedigree chart   Chart visually displaying the lines of your direct ancestors.

personal property   Individual's belongings, excluding land. See *real property*.

portals   Websites that serve as gateways to other places on the web.

power of attorney   Legal document allowing someone else to act on an individual's behalf.

probate   Process of legally establishing the validity of a will.

progenitor   The founder of or first person in a line of descent.

query section   In a genealogical publication, refers to a specific section of the magazine set aside for submitted inquiries.

real property   Refers to land. See also *personal property*.

Register numbering   Genealogical numbering system developed by the New England Historic Genealogical Society for its *Register*; uses an Arabic numeral preceding each child to be carried forward in the compilation; the other children not being carried forward have no arabic number before their names.

repository   Physical location where items are stored for safekeeping. Usually a museum, library, archives, or courthouse.

RSS feed   Really simple syndication. Allows subscribed users to receive automatic updates from websites such as blogs.

SASE (or S.A.S.E.)   Self-addressed stamped envelope.

search engine   Software that searches the web for specific websites and webpages and then compiles them in a database that users can search.

sexton   Caretaker responsible for burials and maintenance of a cemetery.

sic   Means "thus" or "in this manner." Italicized and enclosed in brackets to indicate a word or phrase is rendered as it was in the original. Often used when there is a noted mistake in the original document.

Soundex   Indexing system based on the phonetic sound of the consonants in a surname.

sponsors   People vouching for the suitability of the applicant to be admitted to a lineage society; also those who sponsor a child at baptism.

testator   Person making a will to dispose of his own worldly estate.

topographical map   Detailed, precise description of a place or region with graphic representations of the surface features.

transcribing   The act of faithfully duplicating the exact wording, spelling, and punctuation of an original document.

transcript   Word-for-word exact copy of the text and with all punctuation the same as copied from an original document.

ultimate (ult.)   In dates, refers to the previous month.

URL   Uniform resource locator; the Internet address of a website or webpage.

vault   Steel-lined room designed to prevent destruction of records, such as those found in courthouses.

vertical file   Collection of resource materials often found in libraries; usually pamphlets, letters, or article clippings stored in folders.

vital record   Record for an individual relating to his or her birth, marriage, and death.

webmaster   Person or persons in charge of a website.

will   Legally executed document declaring how a person wishes his or her possessions to be disposed of after death.

witness   In a legal document, one who swears that a signature was obtained in his or her presence.

Y chromosome   The male sex chromosome.

Y-DNA Test   Examination of the Y chromosome for genetic markers.

Relative Resources

Following is a smattering of resources. You'll discover many more, but these will get you started.

BCG Genealogical Standards Manual, The. Washington, D.C.: Board for Certification of Genealogists, 2000.

Bentley, Elizabeth Petty. *Directory of Family Associations.* 4th ed. Baltimore, MD: Genealogical Publishing Co., Inc., 2001.

———. *The Genealogist's Address Book.* 4th ed. Baltimore, MD: Genealogical Publishing Company, 1998. Note: While this book is out of date, it can be useful.

Berry, Ellen Thomas, and David Allen Berry. *Our Quaker Ancestors: Finding Them in Quaker Records.* Baltimore, MD: Genealogical Publishing Company, 1996.

Black's Law Dictionary. 9th ed. St. Paul, MN: West Publishing Company, 2009. Note: the 4th edition covers more obsolete terms.

Bockstruck, Lloyd DeWitt. *Revolutionary War Bounty Land Grants: Awarded by State Governments.* Baltimore, MD: Genealogical Publishing Company, 1996.

Burroughs, Tony. "The Original Soundex Instructions," *National Genealogical Society Quarterly* 89, no. 4 (2001): pp. 287–298.

Carmack, Sharon DeBartolo. *A Genealogist's Guide to Discovering Your Female Ancestors.* Cincinnati, OH: Genealogical Publishing Company, 1998.

———. *Your Guide to Cemetery Research.* Cincinnati, OH: Betterway Books, 2002.

Carr, Peter E. *Guide to Cuban Genealogical Research.* Chicago, IL: Adams Press, 1991.

Census of Pensioners for Revolutionary or Military Services Under the Act for Taking the Sixth Census. Washington, D.C.: Blair and Rives, 1841. Reprint, Baltimore, MD: Southern Book Co., 1954.

Century of Population Growth: From the First Census of the United States to the Twelfth 1790–1900. Washington, D.C.: U.S. Government Printing Office, 1900. Reprint, Orting, WA: Heritage Quest Press, 1989.

Chicago Manual of Style, The. 16th ed. Chicago and London: University of Chicago Press, 2010.

Colletta, John P., Ph.D. *They Came in Ships.* rev. 3rd ed. Salt Lake City, UT: Ancestry, a division of MyFamily.com, Inc., 2002.

Curran, Joan Ferris, CG, Madilyn Coen Crane, and John H. Wray, Ph.D., CG. *Numbering Your Genealogy: Basic Systems, Complex Families and International Kin.* Arlington, VA: National Genealogical Society, 2000.

Drake, Paul J.D. *What Did They Mean by That? A Dictionary of Historical Terms for Genealogists.* Bowie, MD: Heritage Books, Inc., 2009.

Eichholz, Alice, Ph.D., CG, ed. *Ancestry's Red Book.* 3rd ed. Salt Lake City, UT: Ancestry, a division of MyFamily.com, Inc., 2004.

Filby, P. William. *A Bibliography of American County Histories.* Baltimore, MD: Genealogical Publishing Company, 1987.

Gouldrup, Lawrence P., Ph.D. *Writing the Family Narrative.* Salt Lake City, UT: Ancestry Incorporated, 1998.

Greenwood, Val D. *The Researcher's Guide to American Genealogy.* 3rd ed. Baltimore, MD: Genealogical Publishing Company, 2000.

Grundset, Eric, and Steven B. Rhodes. *American Genealogical Research at the DAR.* Washington, D.C.: National Society Daughters of the American Revolution, 2004.

Guide to Genealogical Research in the National Archives. 3rd ed. Washington, D.C.: National Archives, 2000.

Handy Book for Genealogists, The. Logan, UT: Everton Publishers, latest edition published.

Hatcher, Patricia Law, CG. *Producing a Quality Family History,* Salt Lake City, UT: Ancestry Incorporated, 1996.

Hinckley, Kathleen W. *Locating Lost Family Members & Friends: Modern Genealogical Research Techniques for Locating the People of Your Past and Present.* Cincinnati, OH: Betterway Books, 1999.

———. *Your Guide to the Federal Census.* Cincinnati, OH: Betterway Books, 2002.

Index of Revolutionary War Pension Applications in the National Archives. Rev. and enlarged bicentennial ed. Arlington, VA: National Genealogical Society, 1976. Note: this text was originally compiled by Max E. Hoyt and others, and was known as *Hoyt's Index.*

Hone, E. Wade. *Land and Property Research in the United States.* Salt Lake City, UT: Ancestry Incorporated, 2008.

Ingalls, Kay Germain. "Cherchez la Femme! Looking for Female Ancestors." *National Genealogical Society Quarterly* 88, no. 3 (2000): pp. 165-178.

Kirkham, E. Kay. *The Handwriting of American Records for A Period of 300 Years.* Logan, UT: Everton Publishers, Inc., 1981.

Lainhart, Ann S. *State Census Records.* Baltimore, MD: Genealogical Publishing Company, 1992.

Leary, Helen F.M. *North Carolina Research.* 2nd ed. Raleigh, NC: North Carolina Genealogical Society, 1996.

Lynch, Daniel M. *Google Your Family Tree.* Provo, UT: FamilyLink.com, 2008.

Matchette, Robert B., et al. *Guide to Federal Records in the National Archives of the United States.* Washington, D.C.: National Archives and Records Administration, 1995. 3 vol.

Melynk, Marcia D. *Genealogist's Handbook for New England Research.* 4th ed. Boston, MA: New England Historic Genealogical Society, 1999.

Meyerink, Kory L., ed. *Printed Sources: A Guide to Published Genealogical Records.* Salt Lake City, UT: Ancestry Incorporated, 1998.

Military Service Records: A Select Catalog of National Archives Microfilm Publications. Washington, D.C.: National Archives and Service Administration, 1985.

Mills, Elizabeth Shown. *Evidence! Citation and Analysis for the Family Historian.* Baltimore, MD: Genealogical Publishing Company, 1997.

———. *Evidence Explained: Citing History Sources from Artifacts to Cyberspace,* 2nd ed. Baltimore, MD: Genealogical Publishing Company, 2009.

———. ed. *Professional Genealogy: A Manual for Researchers, Writers, Editors, Lecturers, and Librarians.* Baltimore, MD: Genealogical Publishing Company, 2001.

Mokotoff, Gary. "Soundexing and Genealogy," avotaynu.com/soundex.html.

Neagles, James C. *The Library of Congress: A Guide to Genealogical and Historical Research.* Salt Lake City, UT: Ancestry Incorporated, 1990.

———. *U.S. Military Records: A Guide to Federal and State Sources, Colonial America to the Present.* Salt Lake City, UT: Ancestry Incorporated, 1994.

Newman, John J. *American Naturalization Processes and Procedures 1790–1985.* 2nd ed. Indianapolis, IN: Indiana Historical Society, 1998.

PERSI (Periodical Source Index). Fort Wayne, Ind.: Allen County Library Foundation, 1986+. Note: You may also access through subscription to ancestry.com or at libraries subscribing to HeritageQuest.

Rose, Christine. *Courthouse Research for Family Historians: Your Guide to Genealogical Treasures.* San Jose, CA: CR Publications, 2004.

———. *Courthouse Indexes Illustrated.* San Jose, CA: CR Publications, 2006.

———. *Genealogical Proof Standard: Building a Solid Case.* 3rd ed. San Jose, CA: CR Publications, 2009.

———. *Military Bounty Land 1776–1855.* San Jose, CA: CR Publications, 2011.

———. *Military Pension Acts 1776–1858.* San Jose, CA: CR Publications, 2000.

———. *Nicknames Past and Present.* 5th ed. San Jose, CA: CR Publications, 2007.

Rubincam, Milton. *Pitfalls in American Genealogy.* Salt Lake City, UT: Ancestry Incorporated, 1987.

Ryskamp, George R. *Finding Your Hispanic Roots.* Baltimore, MD: Genealogical Publishing Company, 1997.

Schaefer, Christina K. *Guide to Naturalization Records of the United States.* Baltimore, MD: Genealogical Publishing Company, 1997.

Shawker, Thomas H. *Unlocking Your Genetic History: A Step-by-Step Guide to Discovering Your Family's Medical and Genetic Heritage.* Nashville, TN: Rutledge Hill Press, 2004.

Sperry, Kip. *Abbreviations and Acronyms.* 2nd ed. Orem, UT: Ancestry Incorporated, 2003.

———. *Reading Early American Handwriting.* 1998. Reprint, Baltimore, MD: Genealogical Publishing Company, 2008.

Smith, Juliana Szucs Smith. *Ancestry Family Historian's Address Book: The Revised Second Edition.* Salt Lake City, UT: Ancestry Incorporated, 2003.

Smolenyak, Megan Smolenyak, and Ann Turner. *Trace Your Roots with DNA, Using Genetic Tests to Explore Your Family Tree*. Emmaus, PA: Rodale Books, 2004.

Sturdevant, Katherine Scott. *Bringing Your Family History to Life Through Social History*. Cincinnati, OH: Betterway Books, 2000.

Szucs, Loretto Dennis, and Sandra Hargreaves Luebking, eds. *The Source: A Guidebook of American Genealogy*. 3rd ed. Salt Lake City, UT: Ancestry, a division of MyFamily.com, Ind, 2006.

Taylor, Maureen. *Uncovering Your Past Through Family Photographs*. 2nd ed. Cincinnati, OH: Family Tree Books, 2005.

Tepper, Michael. *American Passenger Arrival Records: A Guide to the Records of Immigrants Arriving at American Ports by Sail and Steam*. Updated and enlarged. Baltimore, MD: Genealogical Publishing Company, 1993.

Thorndale, William, and William Dollarhide. *Map Guide to the U.S. Federal Censuses, 1790–1920*. Baltimore, MD: Genealogical Publishing Company, 1987.

U.S. Census Bureau. *Measuring America: The Decennial Censuses from 1790–2000*. Washington, D.C. Also available at census.gov/prod/www/abs/ma.html.

White, Virgil D. *Genealogical Abstracts of Revolutionary War Pension Files*. 4 vols. Waynesboro, TN: The National Historical Publishing Co., 1990.

Computer Programs

AniMap: goldbug.com

DeedMapper: directlinesoftware.com

Clooz: ancestordetective.com

Family Tree Maker: familytreemaker.com

GoogleEarth: google.com/earth/index.html

Personal Ancestral File (PAF): familysearch.org/eng/paf/

Reunion Leister Productions: leisterpro.com

RootsMagic: rootsmagic.com

The Master Genealogist: whollygenes.com

Educational Opportunities

Boston University: professional.bu.edu/programs/genealogy

Brigham Young University Conferences and Workshops: ce.byu.edu/cw/cwgeneal

Federation of Genealogical Societies: fgs.org

Institute of Genealogy and Historical Research: samford.edu/schools/ighr

National Genealogical Society: ngsgenealogy.org

National Institute on Genealogical Research: rootsweb.com/~natgenin

Salt Lake Institute of Genealogy: infouga.org

Worksheets

The worksheets in this appendix are examples of forms you may use to organize your genealogy research. Use the research calendar and the correspondence log to keep a running summary of what you've accomplished.

The research calendar shows you at a glance the records you've checked in reference to a particular surname and problem. Looking at the list, you can determine quickly whether you've searched a particular record and what the results were if you have. The research calendar is meant to be an overview of your work (rather than the place for detailed notes).

The correspondence log is a record of your letter writing. Scanning the list, you can see quickly to whom you've written and his or her response. For more on these two forms, see Chapter 3.

Use the pedigree chart as a sketch of your bloodline. Here's where you record the bare-bones statistics on your ancestors. Once you've filled in the chart with what you know, you can develop a research plan to find the missing information. The family group sheet is the form you use to record more detailed information for the people on your pedigree chart.

Family group sheets are the foundation for organizing the information you collect. Complete at least two family group sheets for each individual on your pedigree chart. On one family group sheet the individual appears as a child and on another as a mother or father. (An individual with more than one marriage should have a family group sheet for each marriage.) Pedigree charts and family group sheets aren't the end products of your research. They're tools to help you organize your findings. See Chapter 4 for more on pedigree charts and family group sheets.

Surname

RESEARCH CALENDAR

Search focus (brief statement of research problem)

Date	Repository/call no.	Source	Findings

Research calendar.

CORRESPONDENCE LOG

Date Sent	To Whom	Request	Reply Date	Results (Positive, Negative, Burned)

Correspondence log.

PEDIGREE CHART

CHART NO. _____

Prepared by _____
Address _____
City and State _____ Zip _____
Date Prepared _____

No. 1 on this Chart is # _____ *on Chart #* _____

1.
Born
Where
Married When
Died
Where

Name of Husband or Wife

2. [Father of No. 1]
Born
Where
Married When
Died
Where

3. [Mother of No. 1]
Born
Where
Died
Where

4. [Father of No.2]
Born
Where
Married When
Died
Where

5. [Mother of No. 2]
Born
Where
Died
Where

6. [Father of No. 3]
Born
Where
Married When
Died
Where

7. [Mother of No. 3]
Born
Where
Died
Where

8.
Born
Where
Married When
Died
Where
Con't on Chart # [Father of No. 4]

9.
Born
Where
Died
Where
Con't on Chart # [Mother of No. 4]

10.
Born
Where
Married When
Died
Where
Con't on Chart # [Father of No. 5]

11.
Born
Where
Died
Where
Con't on Chart # [Mother of No. 5]

12.
Born
Where
Married When
Died
Where
Con't on Chart # [Father of No. 6]

13.
Born
Where
Died
Where
Con't on Chart # [Mother of No. 6]

14.
Born
Where
Married When
Died
Where
Con't on Chart # [Father of No. 7]

15.
Born
Where
Died
Where
Con't on Chart # [Mother of No. 7]

Pedigree chart.

FAMILY GROUP SHEET

NO. _____

HUSBAND [Full name]			SOURCES: Brief listing.	
BORN	AT		No.	
CHR.	AT		No.	
MAR.	AT		No.	
DIED	AT		No.	
BURIED AT			(Complete source citations on reverse)	
FATHER	MOTHER [Maiden Name]			
OTHER WIVES				
RESIDENCES		RELIGION		
OCCUPATION		MILITARY		

[Use separate forms for each marriage]

WIFE [Full maiden name]		
BORN	AT	
CHR.	AT	
DIED	AT	
BURIED AT		
FATHER	MOTHER [Maiden Name]	
OTHER HUSBANDS		

Sex Children		Day-Month-Year		City/Town County State	REF. No.
1.		b.	at		Ref.
Spouse:		m.	at		Ref.
		d.	at		Ref.
2.		b.	at		Ref.
Spouse:		m.	at		Ref.
		d.	at		Ref.
3.		b.	at		Ref.
Spouse:		m.	at		Ref.
		d.	at		Ref.
4.		b.	at		Ref.
Spouse:		m .	at		Ref.
		d.	at		Ref.
5.		b.	at		Ref.
Spouse:		m.	at		Ref.
		d.	at		Ref.
6.		b.	at		Ref.
Spouse:		m.	at		Ref.
		d.	at		Ref.
7.		b.	at		Ref.
Spouse:		m.	at		Ref.
		d.	at		Ref.
8.		b.	at		Ref.
Spouse:		m.	at		Ref.
		d.	at		Ref.
9.		b.	at		Ref.
Spouse:		m.	at		Ref.
		d.	at		Ref.
10.		b.	at		Ref.
Spouse:		m.	at		Ref.
		d.	at		Ref.

PREPARED BY	OTHER MAR. OF CHILDREN
Address	Use reverse for additional marriages
City and State	Zip
Date Prepared:	

b.=born   m.=married   d.=death   ch.=christening   List references at top and use ref. numbers on items. Use reverse if necessary.

Family group sheet.

Here's a checklist of questions for interviewing relatives. You'll think of others as you devise your own lists for each interview you conduct. See Chapters 2 and 20 for more on interview techniques.

You need to know names, dates, and locations, but if you bombard your relatives with those questions, they may become flustered—or worse yet, they may lose interest. Try to fit those questions in among the ones that trigger reminiscences. Much of the information you'll find in your search will pertain to your male ancestors. During interviews, make a special effort to elicit information about your female ancestors, too.

It isn't likely you'll have time to ask all your questions, so decide which ones are most important to you, but don't be too rigid. Through the natural course of the conversation, you may get answers to questions you didn't think to ask or were hesitant to ask. The answer to one question could lead to something you'll want to explore further. The following questions will get you started with your own list; you'll think of others. The list refers to males; adapt the same list for females.

- What was Grandpa's full name? Was he named for anyone else? What was he called? How did he get his nickname?

- When and where was he born? Was it a small town or a farming community? Do you know what brought the family there?

- What was his father's name?

- What was his mother's maiden name?

- Did Grandpa have brothers and sisters? What were their names? When and where were they born? Who did they marry?

- Where were the various towns in which the family lived? Did they like moving around? Were they following the available work or were they just adventurous? Did other relatives move with them?

- What were the names of Grandpa's aunts and uncles? Did they live nearby? Were there often large family gatherings?

- What did Grandpa look like? Did he have a beard? Did he have a small frame or a large frame?

- Does anyone have any photos of him? Did he resemble anyone else in the family?

- What did Grandpa do to earn a living? Did he like it? Did any of the children follow in his footsteps?

- Where is he buried? Have you ever been there? Is there a tombstone?

- Did Grandpa ever serve in a war? Does anyone have his uniform or other paraphernalia from the war?

- What church did he attend? Did he attend with Grandma? Was he active in the church?

- Did he have a trade or a hobby? Did he like gardening? Did he have any pets?

- Did he own land? Was it cultivated for crops? Did he like working on it?

- What was his nationality? Did he speak with an accent?

- Does anyone in the family have any of his possessions that were handed down? Do you know the stories about the items?

- Tell me about your holiday dinners. Were birthdays celebrated? What kinds of gifts did you give and receive?

- What did the family do for recreation? Did anyone have a special hobby?

- How was the family's health? Robust or sickly? Allergies? Home remedies?

- Did the family embrace new ideas and technologies? When did they get their first automobile and electricity? Did they have indoor plumbing?

- What is the origin of the family expressions? Did the family have any traditions?

- Were they involved in politics or civics? Did they belong to any clubs?

- What were their attitudes toward religion, race, and liquor?

- Did anyone play a musical instrument or have artistic talent?

- Was anyone athletic?

- Is there a family trait that seems pervasive?

- How did you meet your spouse? How did you decide on a wedding date?

- What was your school like—one room in the country or a brick building in the city? Were you a good student?

- Who were your best friends, and what did you do together?

- What was your favorite age? Why?

- Did you get along well with your siblings and/or your cousins?

Census Forms

The following pages contain forms for the 1800–1940 censuses. You may photocopy these forms and use them to note information you've extracted from microfilm.

Page	Head of Family	Free White Males					Free White Females					All Others	Slaves	Remarks
		Under 10	10–16	16–26	26–45	45 & Over	Under 10	10–16	16–26	26–45	45 & Over			

State ____ County ____ **1800–1810 CENSUS — UNITED STATES** City ____ Call No. ____

1800–1810 Census Form.

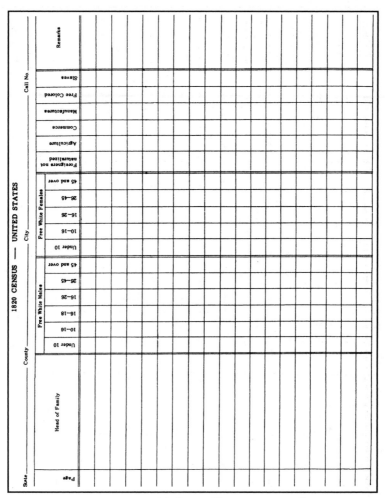

1820 Census Form.

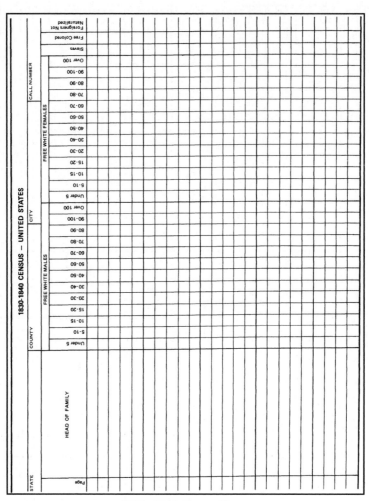

1830–1840 Census Form.

Page	Dwelling Number	Family Number	NAMES	Age	Sex	Color	OCCUPATION, ETC.	Value—Real Estate	BIRTHPLACE	Married Within Year	School Within Year	Can't Read or Write	Enumeration Date	REMARKS

1850 CENSUS – UNITED STATES

STATE _____ COUNTY _____ TOWN/TOWNSHIP _____ CALL NUMBER _____

1850 Census Form.

1860 Census Form.

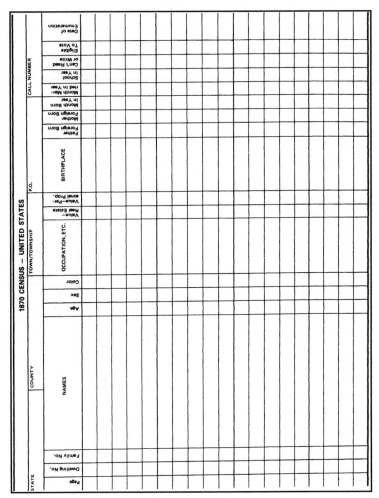

1870 Census Form.

1880 CENSUS – UNITED STATES

STATE

COUNTY

TOWN/TOWNSHIP

CALL NUMBER

Page	Dwelling Number	Family Number	NAMES	Color	Sex	Age Prior to June 1st	Month of Birth if Born in Census Year	Relationship to Head of House	Single	Married	Widowed	Divorced	Married in Census Year	Occupation	Miscellaneous Information	Cannot Read or Write	Place of Birth	Place of Birth of Father	Place of Birth of Mother	Enumeration Date

1880 Census Form.

1900 CENSUS – UNITED STATES

MICROFILM ROLL NUMBER

STATE — TOWN/TOWNSHIP — SUPV. DIST. NO. — SHEET NUMBER

COUNTY — CALL NUMBER — ENUM. DIST. NO. — PAGE NUMBER

DATE

LOCATION
- Street
- House Number
- Dwelling Number
- Family Number

NAME of each person whose place of abode on June 1, 1900, was in this family

Relation to Head of family

PERSONAL DESCRIPTION
- Color
- Sex
- Month of birth
- Year of birth
- Age
- Single, married, widowed, divorced
- Number of years married
- Mother of how many children
- Number of these children living

NATIVITY
- Place of birth
- Place of birth of father
- Place of birth of mother

CITIZENSHIP
- Year of immigration to United States
- No. of years in U.S.
- Naturalization

OCCUPATION
- Type
- Number of months not employed

EDUCATION
- Attended school (months)
- Can read
- Can write
- Can speak English

- Home owned or rented
- Home owned free or mortgaged
- Farm or house

1900 Census Form.

1910 CENSUS – UNITED STATES

STATE _____ COUNTY _____ TOWN/TOWNSHIP _____ ENUM. DIST. NO. _____ PAGE NO. _____

| LOCATION | | | | NAME | RELATION TO HEAD OF HOUSE | PERSONAL DESCRIPTION | | | | | | | NATIVITY | | |
STREET NAME	HOUSE NUMBER	VISITATION NUMBER	FAMILY NUMBER	OF EACH PERSON WHOSE PLACE OF ABODE ON APRIL 15, 1910, WAS IN THIS FAMILY		SEX	RACE	AGE	SINGLE/MARRIED/ WIDOWED/DIVORCED	NUMBER OF YEARS PRESENT MARRIAGE	NO. OF CHILDREN BORN THIS MOTHER	NUMBER OF THESE CHILDREN LIVING	PLACE OF BIRTH OF THIS PERSON	PLACE OF BIRTH OF FATHER	PLACE OF BIRTH OF MOTHER

1910 CENSUS – UNITED STATES (CONTINUED)

STATE | COUNTY | TOWN/TOWNSHIP | ENUM. DIST. NO. | PAGE NO.

NAME — OF EACH PERSON WHOSE PLACE OF ABODE ON APRIL 15, 1910, WAS IN THIS FAMILY (FROM OTHER SIDE OF THIS FORM)

CITIZENSHIP
- YEAR OF IMMIGRATION TO U.S.
- NATURALIZED/ALIEN
- NATIVE LANGUAGE

OCCUPATION
- TRADE OR PROFESSION
- NATURE OF BUSINESS
- EMPLOYER/EMPLOYEE/SELF-EMPLOYED
- IF EMPLOYEE: UNEMPLOYED, WEEKS OUT OF WORK IN 1909

EDUCATION
- ABLE TO READ
- ABLE TO WRITE
- ATTENDED SCHOOL SINCE SEPT 1, 1909

HOME OWNERSHIP
- OWNED/RENTED
- OWNED FREE/MORTGAGED
- FARM/HOUSE
- NO. OF FARM SCHEDULE

- UNION/CONFEDERATE VETERAN
- BLIND
- DEAF AND DUMB

1910 Census Form.

1920 CENSUS — UNITED STATES

STATE

COUNTY

TOWNSHIP OR OTHER COUNTY DIVISION

NAME OF INSTITUTION

NAME OF INCORPORATED PLACE

ENUMERATED BY ME ON THE _____ DAY OF _____, 1920.

SUPERVISOR'S DISTRICT #

ENUMERATION DISTRICT #

SHEET NO.

WARD OF CITY

ENUMERATOR

PLACE OF ABODE				NAME	RELATION	TENURE		PERSONAL DESCRIPTION					CITIZENSHIP			EDUCATION		
STREET, AVENUE, ETC.	HOUSE NUMBER OR FARM	NUMBER OF DWELLING HOUSE (VISITATION ORDER)	NUMBER OF FAMILY (VISITATION ORDER)	OF EACH PERSON WHOSE PLACE OF ABODE ON JANUARY 1, 1920, WAS IN THIS FAMILY	RELATIONSHIP TO HEAD OF HOUSEHOLD	HOME OWNED OR RENTED	IF OWNED, FREE OR MORTGAGED	SEX	COLOR OR RACE	AGE AT LAST BIRTHDAY	SINGLE, MARRIED WIDOWED, OR DIVORCED	YEAR OF IMMIGRATION TO U.S.	NATURALIZED OR ALIEN	IF NATURALIZED, YEAR OF NATURALIZATION	ATTENDED SCHOOL ANYTIME SINCE SEPT. 1, 1919	ABLE TO READ	ABLE TO WRITE
1	2	3	4	5	6	7	8	9	10	11	12	13	14	15	16	17	18

1920 CENSUS — UNITED STATES

STATE

COUNTY

TOWNSHIP OR OTHER COUNTY DIVISION

NAME OF INSTITUTION

SUPERVISOR'S DISTRICT #

ENUMERATION DISTRICT #

SHEET NO.

NAME OF INCORPORATED PLACE

WARD OF CITY

ENUMERATED BY ME ON THE _____ DAY OF _____, 1920

ENUMERATOR

NAME

OF EACH PERSON WHOSE PLACE OF ABODE ON JANUARY 1, 1920, WAS IN THIS FAMILY
(from other side of form).

5

NATIVITY AND MOTHER TONGUE

PLACE OF BIRTH OF EACH PERSON AND PARENTS OF EACH PERSON ENUMERATED. IF BORN IN U.S., GIVE STATE OR TERRITORY. IF FOREIGN BIRTH, GIVE THE PLACE OF BIRTH, AND, IN ADDITION, THE MOTHER TONGUE.

PERSON — PLACE OF BIRTH 19

PERSON — MOTHER TONGUE 20

FATHER — PLACE OF BIRTH 21

FATHER — MOTHER TONGUE 22

MOTHER — PLACE OF BIRTH 23

MOTHER — MOTHER TONGUE 24

ABLE TO SPEAK ENGLISH 25

OCCUPATION

TRADE, PROFESSION, OR PARTICULAR KIND OF WORK DONE. 26

INDUSTRY, BUSINESS, OR ESTABLISHMENT IN WHICH AT WORK. 27

EMPLOYER, SALARY OR WAGE WORKER, OR WORKING ON OWN ACCOUNT. 28

NUMBER OF FARM SCHEDULE 29

1920 Census Form.

1930 Census United States

State ——— County ——— Township or Other County Division ——— Name of Institution ———
Name of Incorporated Place ——— Enumeration Date ——— 1930 Enumerator ———
Ward of City ——— Supervisor's District ——— Enumerator's District ——— Sheet No. ———

1930 Census Form.

1940 Federal Census

STATE

COUNTY

TOWNSHIP OR OTHER DIVISION OF COUNTY

INCORPORATED PLACE

WARD OF CITY

ENUMERATION DISTRICT NO.

SUPERVISOR'S DISTRICT NO.

ENUMERATED BY ME ON _____, 1940

INSTITUTION

SHEET NO.

_____, ENUMERATOR.

BLOCK NO.

UNINCORPORATED PLACE

LOCATION

Line No.

Street, Avenue, road, etc.

House Number

HOUSEHOLD DATA

No. of Household in order of visitation

Home owned (O) or rented (R)

Value of home, if owned, or monthly rental if rented

Farm? (Yes or No)

NAME

Name of each person whose usual place of residence on April 1, 1940, was in this household

BE SURE TO INCLUDE:
1. Persons temporarily absent from household. Write "Ab" after names of such persons.
2. Children under 1 year of age. Write "Infant" if child has not been given a first name.

Enter other name of persons furnishing information.

RELATION

Relationship of this person to the head of the household, as wife, daughter, father, mother-in-law, grandson, lodger's wife, servant, hired hand, etc.

PERSONAL DESCRIPTION

Sex

Color or Race

Age at Last Birthday

Marital Status

EDUCATION

Attended school or college any time since March 1, 1940?

Highest grade of school completed

PLACE OF BIRTH

If born in U.S. give state, territory or possession.

If foreign born, give country in which birthplace was situated on Jan. 1, 1937.

Distinguish Canada-French from Canada-English and Irish Free State from Northern Ireland.

CITIZENSHIP

Citizenship of the foreign born

RESIDENCE, APRIL 1, 1935

In what place did this person live on April 1, 1935? For a person who lived in a different place, enter city or town, county, and State

City, town, or village having 2,500 or more inhabitants. If less, enter "R"

County

State (or Territory or foreign country)

On a Farm? (Y or N)

PERSONS 14 YEARS OLD AND OVER — EMPLOYMENT STATUS

Was this person AT WORK for pay or profit in private or nonemergency work during week of March 24-30, 1940? (Y or N)

If not, was he at work on, or assigned to, public EMERGENCY WORK (WPA, NYA, CCC, etc.) during week of March 24-30? (Y or N)

Was this person SEEKING WORK? (Y or N)

If not seeking work, did he HAVE A JOB, business, etc.? (Y or N)

If neither at work nor assigned to public emergency work ("No" in cols. 21 &22)

Indicate whether engaged in home housework (H), in school (S), unable to work (U), or other (Ot).

For persons answering "No" to questions 21-24

Number of hours worked during week of March 24-30, 1940.

Duration of unemployment up to March 30, 1940 — in weeks

If at private or nonemergency Govt. work. "Yes" in col. 21 or 24

OCCUPATION, INDUSTRY, AND CLASS OF WORKER

For a person at work, assigned to public emergency work, or with a job ("Yes" in col. 21, 22, or 24), enter present occupation, industry, and class of worker.
For a person seeking work ("Yes" in col. 23), (a) if he has previous work experience, enter last occupation, industry, and class of worker; or (b) if he does not have previous work experience, enter "New worker" in Col. 28, and leave Cols. 29-30 blank.

OCCUPATION

Trade, profession, or particular kind of work, as —
Frame spinner
Salesman
Laborer
Rivet heater
Music teacher

INDUSTRY

Industry or business, as —
Cotton mill
Retail grocery
Farm
Shipyard
Public school

Class of Worker

CODE (leave blank)

Number of weeks worked full-time in 1939

INCOME IN 1939 (12 months ending Dec. 31, 1939.)

Amount of money, wages or salary received (including commission)

Did this person receive income of $50 or more from sources other than money wages or salary (Y or N)

Number of Farm Schedule

Line No.

National Archives and Records Administration

NARA's web site is http://www.archives.gov

NA 14129 (6-09)

SUPPLEMENTARY QUESTIONS

For Persons Enumerated on Lines 14 and 29.

FOR PERSONS OF ALL AGES

FOR PERSONS 14 YEARS OLD AND OVER

FOR ALL WOMEN WHO ARE OR HAVE BEEN MARRIED

Line No.	Name	PLACE OF BIRTH OF FATHER AND MOTHER. If born in U.S. give state, territory, or possession. If foreign born, give country in which birthplace was situated on Jan. 1, 1937. Distinguish Canada-French from Canada-English and Irish Free State from Northern Ireland.		MOTHER TONGUE. Language spoken in home in earliest childhood.		VETERANS. Is this person a veteran of the United States military forces; or the wife, widow, or under-18-year-old child of a veteran?					SOCIAL SECURITY			USUAL OCCUPATION, INDUSTRY, AND CLASS OF WORKER. Enter that occupation which the person regards as his usual occupation and at which he is physically able to work. If the person is unable to determine this, enter that occupation at which he has worked longest during the past 10 years and at which he is physically able to work. Enter also usual industry and usual class of worker.					
		Father	Mother		CODE (leave blank)	If so enter "Yes"	If child, is veteran-father dead? (Y or N)	War or Military Service	CODE (leave blank)	Does this person have a Federal Social Security Number? (Yes or No)	Where deductions for Federal Old-Age Insurance or Railroad Retirement made from wages/salary in 1939? Census. (No)	If so, were deductions made from all, ½ or more, part but less than ½, of wages or salary?	Usual Occupation	Usual Industry	Usual class of worker	CODE (leave blank)	Has this woman been married more than once? (Yes or No)	Age at first marriage.	Number of children ever born. (Do not include stillbirths.)
	22	36	37		39	39	40	41	1	42	43	44	42	44	45	2	47	48	50
14																			
29																			

1940 Census Form.

Index